P9-EIG-364

The Art of

ENAMELING

Pewter cup with gold cloisonné. Height: about 6½ inches. (Collection of Mrs. Fred Rubman)

Margaret Seeler

The Art of ENAMELING

How To Shape Precious Metal and Decorate It with Cloisonné, Champlevé, Plique-à-Jour, Mercury Gilding and Other Fine Techniques

GALAHAD BOOKS · NEW YORK CITY

My sincere thanks to Corinne Reinhold, who inspired the making of this book; to Martha Overkamp, Leon La Plante, and Robert Lentini who took the pictures; and to my son, Peter Zeitner, who helped me with the manuscript. I could not have done it without you.

 Manufactured in the United States of America.
Designed by Rosa Delia Vasquez

Library of Congress Catalog Card Number: 74-75598
ISBN 0-88365-239-0

Published by arrangements with Van Nostrand Reinhold Company

CONTENTS

Introduction

Many books have been written about enameling. Each author has his own way of working with this versatile medium. Enamel offers the artist-craftsman the freedom of painting, graphic arts, and even sculpture. This is not a new challenge. Through the ages there have been different approaches to enameling, all of them leading to marvelous results, when the execution was right and the artist sincere.

I will mention some of the old techniques – not merely by their French names which signify a specialty, but in order to show what they can teach us and how they can inspire us. We may have less training in workmanship than the medieval masters, but we have new technical facilities for achieving good and lasting results.

All information about enameling was gained from my own experiences and mistakes. I am grateful for what I have learned at the *Vereinigte Staatsschulen fuer freie und Angewandte Kunst* (Academy of Fine and Applied Arts) of Berlin, Germany. But it was here in the United States that I became truly an enameler.

Much of the work shown would never have been completed without the cooperation of the pewtersmith, Frances Felten. One craftsman alone cannot possibly master the skills of several media. Frances Felten's great knowledge and ability combined with my own efforts were a most rewarding experience. I do thank her.

Silver box with grisaille. Self-portrait done from life. 2½ inches long. (Collection of Hans Zeitner)

Adam and Eve. Plaque in silver cloisonné enamel. Fired in a very small hotplate kiln and made with only a few tools. (First prize, All-American Show in Wichita, Kansas, 1959. Owned by Wichita Art Association)

Enamel and precious metal will last. In a thousand years enamel work will be as it is today, and it will reveal a great deal about our skills and our thoughts. What better medium could we find for objects which try to bring eternal values to our understanding . . . or to preserve a smile over the ages?

Photography took from the artist the necessity for depicting the realistic appearance of men and objects. There seemed to be no more need to draw precisely. One field, however, that remains open to the artist and that photography can never invade is the expression of our inner visions.

To avoid stumbling over poor execution and to build a treasury of inner pictures, we must study, sketch, and observe life and nature and be able to reduce their outward forms to intense, visual expression in our work. Cloisonné is an almost ideal technique for the artist who wishes to force himself to this simplicity, the most difficult task.

To me, art is a language beyond time, space, and nationality. It can be understood by those whom we may never meet. Sincerity, honest work, and self-restraint will go much further than short cuts and all the tools money can buy. Even a beginner is able to work well and in good taste with a few basic tools, and every step will carry with it the joy of achievement.

And now let's get to work . . .

1.

Equipment

The work area.

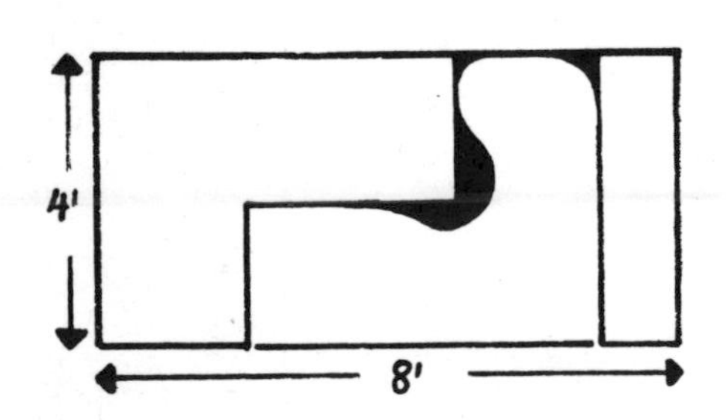

The whole work area is cut out of one piece of plywood measuring 4′ x 8′ x 1″.

Our work area has one place for metalsmithing and one for enameling. I hope that you master a few basic skills of silversmithing, like sawing, soldering, and the use of the hammer to some extent, as well as buffing and polishing of metal. If not, you can still work with precast shapes, or you may need a silversmith or pewtersmith to set your enamels later.

If the shop has running water, that is fine. My sketch gives some idea of a work-place which serves the purpose well. One small area should be reserved for sifting enamel and working with stencils; two square feet would be sufficient. Tack paper against the wall for protection from sprayed gum, and place a sheet of clean, white paper on the table. Have a stack of glossy paper, about the size of a magazine, at hand over which to sift. The enamel which doesn't settle on the work will be collected on the top sheet of paper and can be poured back into its jar. Use a fresh piece of paper for each color as you can never separate the colors again.

Almost all the pieces shown in this book have been fired in a good small kiln, only 8½ x 8½ x 4 inches inside. Even a much smaller and rather primitive hotplate-kiln with a glass cover can do a good job. Cover the glass top with heavy-duty aluminum foil to reflect the heat. Cut a small hole into the foil so that you can see when the enamel is molten and the surface shiny and smooth. It will be glowing red by then.

If your kiln is a small one and you work on a large piece, the heat will not fuse your enamel evenly. While the portion nearest the door is not yet mature, the portion deep inside the kiln may be almost overfired.

Attach a piece of asbestos to the top of a Lazy Susan and keep this gadget standing on top of your kiln. It should be large enough to hold the trivet with the enamel.

When the one portion of your enamel is well fired, take it out, place it on the Lazy Susan, give it a quick 180° turn and in seconds you can put the piece back in the kiln the other way around.

It is even possible to enamel without a kiln, using a blowtorch held underneath the surface to be enameled. With precious metal this technique is helpful when a sample of a certain enamel over a certain metal is needed quickly. A small piece of the metal and a few specks of the color, applied either wet or dry with a spatula, will begin to glow immediately over a blowtorch. Since such samples have no counterenamel they will not last, but they do tell us what we need to know.

If the enameler has to repeat the same piece many times in the same quality, a pyrometer is essential. The exact temperature and firing time must be known.

The artist-craftsman who does the one-of-a-kind type of work needs no pyrometer, though of course it is helpful to have one. A big bakery has to be able to produce the same quality in hundreds of loaves of bread every day, but the gourmet cook knows his dough and his oven so well that he can control the process by instinct. If you are able to estimate the temperature by the color, you are independent and can be sure even with modest tools.

Small hotplate-kiln with foil reflector.

Following is a temperature-color guide:

1200 to 1300	degrees	Fahrenheit	— *dark red*
1300 to 1400	"	"	— *dark to cherry-red*
1400 to 1500	"	"	— *cherry-red*
1500 to 1550	"	"	— *light cherry-red*
1550 to 1600	"	"	— *cherry to orange-red*
1600 to 1650	"	"	— *orange-yellow*

Dark red will not fire enamels to maturity, but it is hot enough to fire grisaille, china-paints, pure gold, and pure silver.

Cherry-red will anneal metal. It will, if you watch it, do what the darker red does: let the piece become just dark red. The enamel will melt to a surface which might be called "orange peel," still rough. We start with such a surface for a three-dimensional effect.

Light cherry-red is good for fast and high firing except when working with gold and copper. These may take even a *cherry to orange-red* for the last firing, which brings out brilliance in the work.

Orange-red is close to the danger-line. There are situations when we may need such a fast and high fire, but never with silver.

Orange-yellow is good for handsoldering in the kiln.

The old-fashioned goldsmith's table is still the best for work with precious metals. A piece of metal tacked to the wood makes a good fireproof surface.

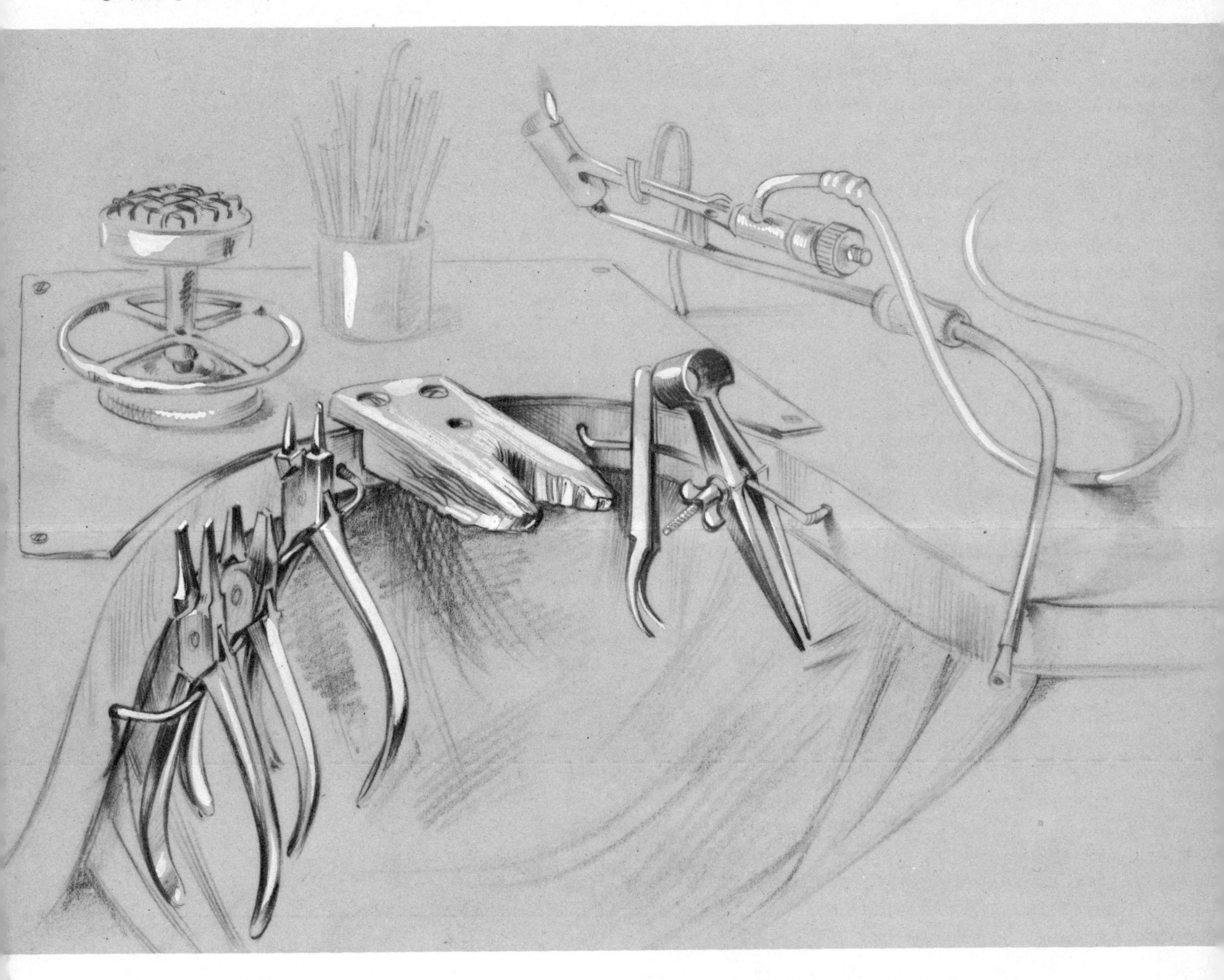

TOOLS FOR METALWORK

You will need the following:

1. Three pairs of good pliers: flat-nosed, half-round, and round-nosed.
2. One pair of good sturdy tweezers and one of self-locking cross-lock tweezers.
3. One light planishing hammer and a small watchmaker's rivet-hammer.
4. A saw-frame and the best sawblades #3/0 and 4/0.

5. Files: (cross-cut) flat, half-round, and round; slitting and crossing. One larger square and one equaling as well would be good to have.

6. Wet-and-dry emery paper, medium or coarse. (A silicon carbide on waterproof paper, it does not crumble and leave particles on the work, and is used for fine finishing.)

7. An electric motor for polishing and possibly a flexible shaft with drills, abrasive wheels, buffs, and diamond-impregnated points and wheels for repairs and polishing.

8. If gas is available, use a mouth blowtorch for fine work. Your own lungs provide the air pressure and you can regulate it instantly. These torches are available for use with propane gas as well. Such a mouth-torch gives you a wide range of flames, from a soft and quite large one for warming, to finely pointed and very hot for soldering in small areas, or small and gentle for attaching findings with pewter-solder, for example.

9. You will need hard and medium silver-solder, and of course gold-solder too, if you plan to work with gold, as well as a scraper. I am referring here to a beginner, not to an advanced metalsmith.

10. Get a covered Pyrex dish for Sparex or sulfuric acid (called "pickle" by the men from the trade). It need not be large if you intend to work with small pieces of precious metal.

11. One very important tool is a glass brush, which, used under running water, cleans and slightly polishes without leaving a residue.

12. Finally, to protect solder-joints in repeated fire, use ochre.

Your tools must feel good in your hands. When they are still new, they feel like strangers. Go over their sharp edges with emery paper, and buff and polish them until they feel like part of yourself. If your tools are rusty and neglected, reshape them and after a treatment with oil and a vigorous buffing they will become your favorites.

The old-fashioned goldsmith's table (it can be merely a table top clamped to a regular table) has great merits: A piece of leather or leatherette is fastened beneath the curved cutout. Whatever falls down — bits of gold and silver, precious stones, small objects and tools — is caught in this leather. It is amazing how much precious metal accumulates in a year's work.

The table top ought to be three feet high; or so that you can sit comfortably in the cutout and have the work right before your eyes. Keep your eyes shaded, but the work area should be well lighted.

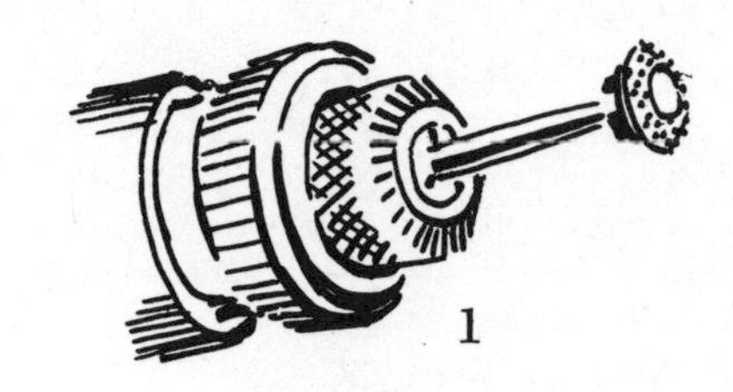

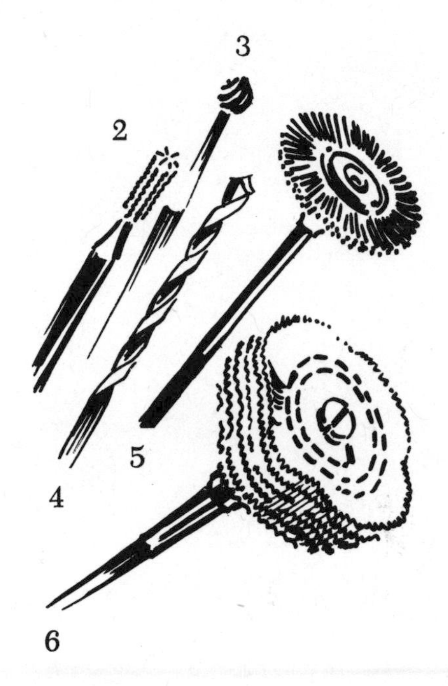

(1) Small diamond-impregnated wheel, ideal for repairs of enamel. (2) Ball burr. (3) Cylinder square burr. (4) Twist drill. (5) Bristle brush for buffing. (6) Muslin buff for polishing.

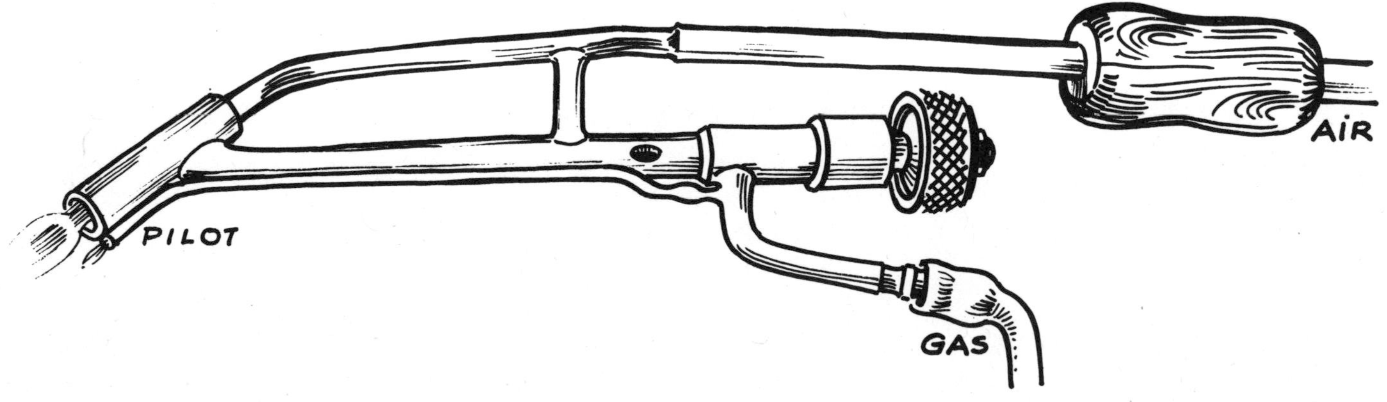

Mouth blowtorch for propane gas (bottle gas). With such a mouthtorch, work is possible wherever bottled gas is available.

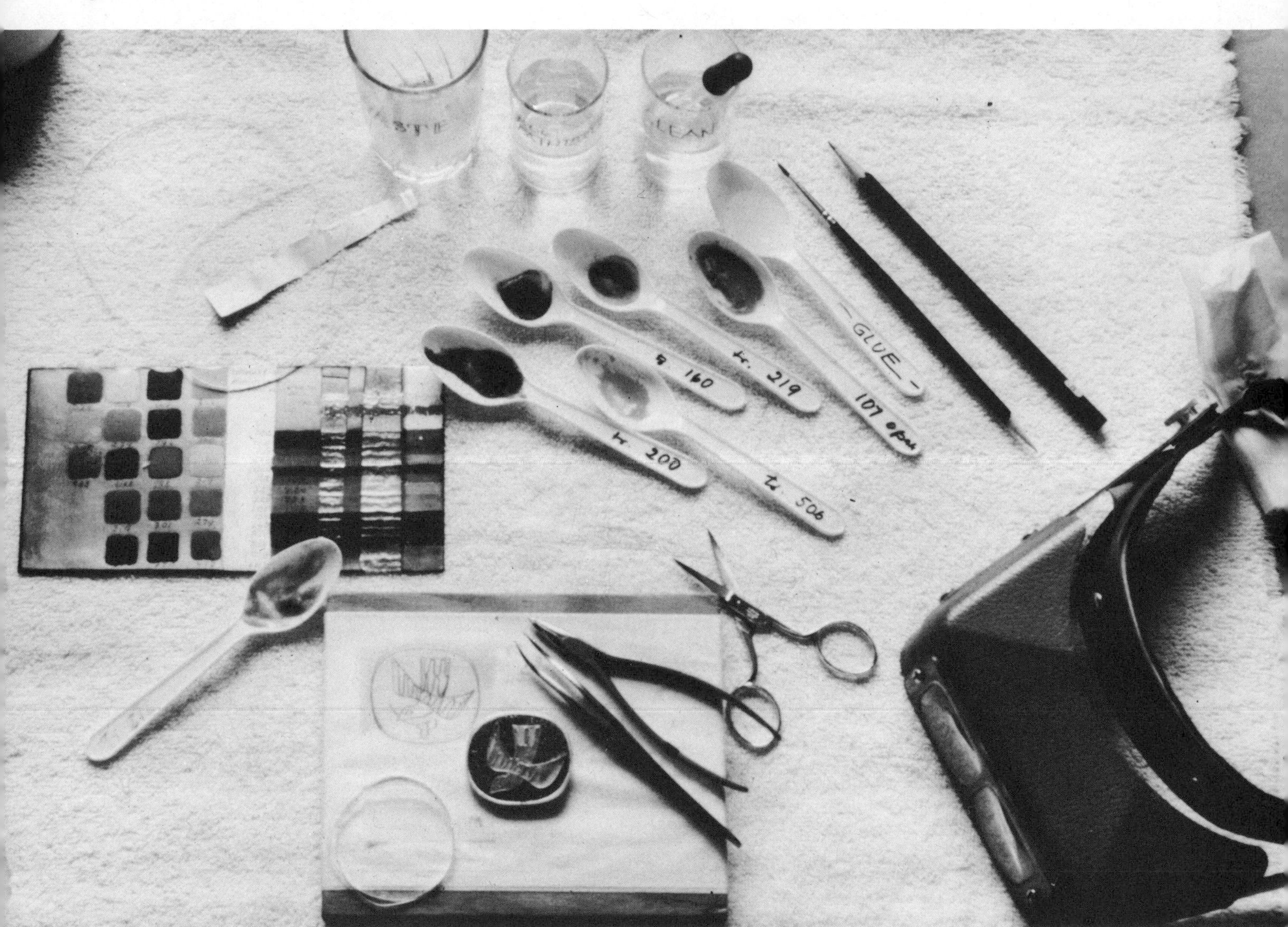

An array of tools and materials for enameling. There are three jars, one for waste, one for cleaning brushes, one to hold clean water for the enamels. Washed enamels are in small white plastic spoons. The number of the enamel and whether it is transparent or opaque is written on the handle of each spoon. At left are small color samples fired on copper. A Kleenex-covered wooden work block gives easy access to the piece from all sides. At right is a magnifying-glass headset.

TOOLS FOR ENAMELING

Enameling can be done almost anywhere, but the place must be immaculate and free of dust. You need electricity and a large board (2 ft. x 3 ft.) covered with white paper. When you are not using it, keep it covered with clean linen or plastic. I find it practical to have a white terrycloth towel under my tools and colors while I wetpack enamels or bend cloisonné wires. It soaks up the moisture from the paintbrushes, is soft, and can easily be replaced.

1. The small pliers for bending cloisonné wires should be watchmaker's pliers of the best quality. File them down in such a manner that you will need only *one* tool for every possible shape to be bent: one jaw should remain flat inside, but the sides should be filed to give sharp 90° angles. The other jaw should be filed to a tapered, round shape. The jaws must touch when the pliers are closed. Buff and polish the pliers with as much care as you would a piece of fine jewelry. They must feel like a part of your hand and turn easily to any necessary position. The rectangular tip is for bending sharp corners, the round tip for making small circles and to hold the wire while making gentle curves.

2. One pair of small watchmaker's tweezers.

3. A fine pair of small scissors, which must cut at the very tip.

4. One good sable brush, either #0 or #1.

5. Small white plastic spoons.

6. Three glasses for water.

7. Klyrfire or tragacanth (binders which do not decompose for some time and fire away completely).

8. Several pieces of mica.

9. Trivets to hold the piece in the kiln, and tongs to place the work in the kiln and to remove it. (See page 24 for more about trivets.)

10. A couple of sieves, which you can buy or make yourself.

11. An old surgical tool with a ripple handle, which you might be able to get from your dentist. This tool is passed over the edge of the piece after wetpacking. The vibration smoothes the enamel and lets the moisture come to the surface where it can easily be absorbed with a Kleenex.

12. Distilled water, though clean well water will do.

13. If you plan to do very fine work, you should have a magnifying glass.

14. A kiln and enamel colors.

15. Lastly, it is most comfortable to place your piece on a small wooden block which you can turn around, thus reaching the work from all sides without moving it from its rest.

(1) The small pliers are filed so that every possible shape of cloisonné wire can be bent. (2) The watchmaker's tweezers must be kept immaculate. Never use them when soldering. (3) The rippled handle of an old surgical tool evens a wetpacked surface when you slide it along the rim of the enamel. (4) A pin, bent at an angle and stuck into the end of a pencil, is a perfectly good tool for packing enamel tightly. It is called a spreader.

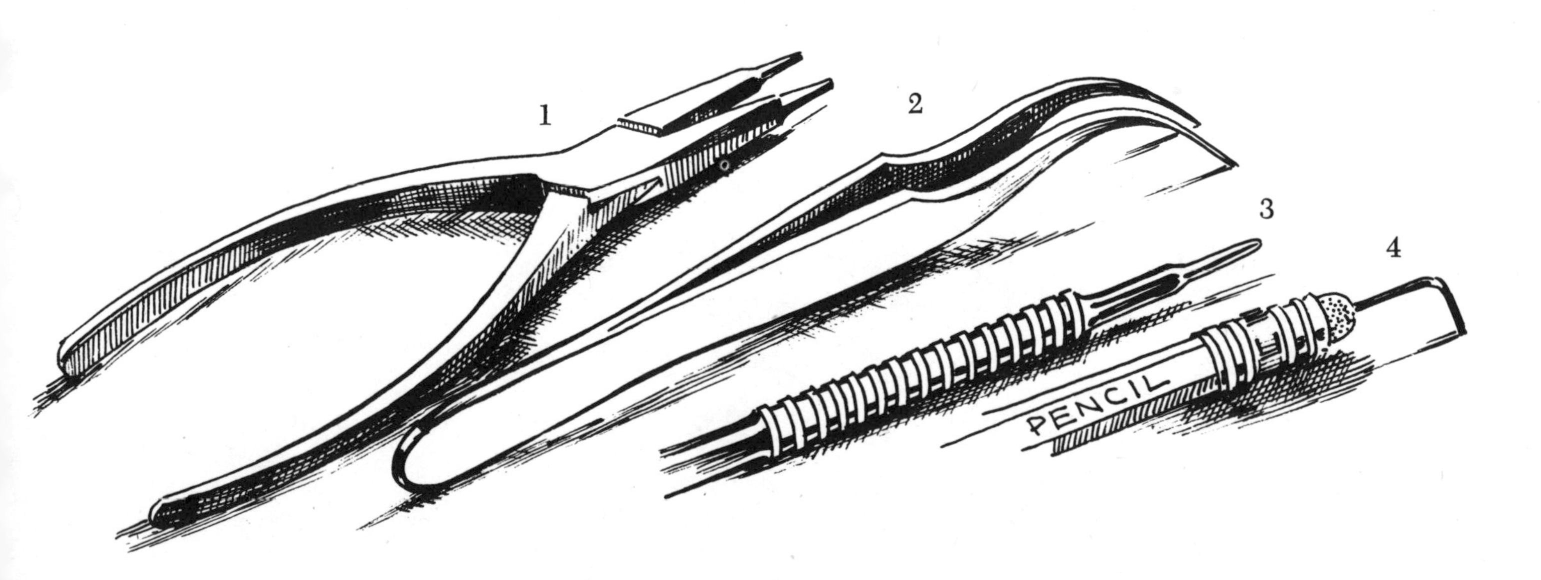

2.

Enamel

The word "enamel" has two different meanings in our work: one refers to the finished piece, the other to the colored glass which is used in making "enamels," those finished pieces of enamel on gold, silver, copper, or certain types of bronze. Enamel is a type of glass, colored by metal oxides (gold in the reds) with approximately the same melting point. It can be coarse-ground or very fine and is applied to the metal either with water or sifted dry over the surface. Then it is fired at about 1500° Fahrenheit. At this point it melts and adheres to the metal.

Enamel adds color to fine metalwork and every craftsman who becomes fascinated with this medium can find new ways to use it. This has happened for a couple of thousand years in many parts of the world.

DIFFERENT PROPERTIES OF ENAMELS

Fluxes

Flux is a first transparent coat of enamel over the metal. It is also used as a last filler for light transparents or as a last glassy cover over a finished piece. There are special fluxes for gold, fine silver, sterling silver, and copper.

To explain the need for so many kinds of flux, let us see where each should be used and also where no flux at all is necessary.

Flux for copper: Transparent colors over copper change as thin water colors would on dark-red paper. However, you will retain a beautiful

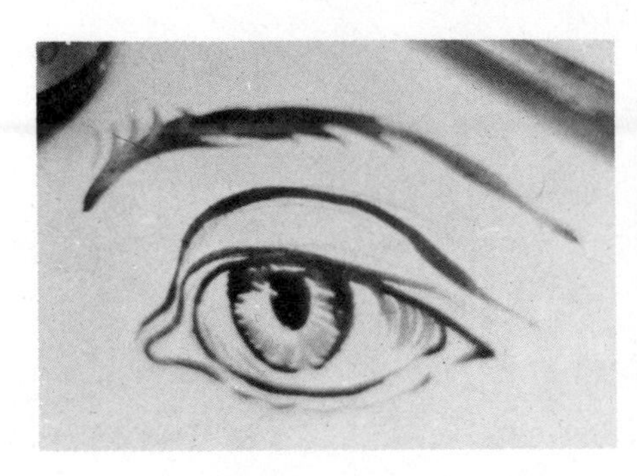

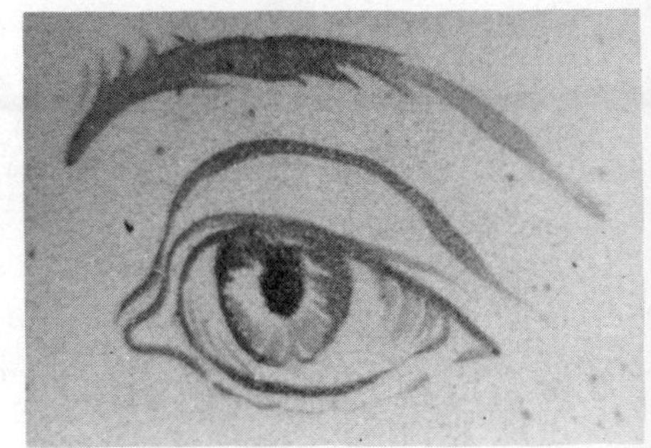

A sample of painting with black Underglaze D directly on copper; next to it, the same sample *fired,* under a coat of flux for copper.

golden color on copper if you sift a coat of copperflux over the bright and grease-free metal and fire this quite high (light cherry-red). Use plenty of flux around the edges; this is where it tends to burn away. If the firing was too short or too low in temperature, the flux will still be red. Sift another thin coat over the piece and refire until it glows red. Flux will not crack, even though the piece is not yet counterenameled. Flux is quite elastic; if the piece has warped you can bend it back to shape. Tiny cracks might occur, but they disappear in the next firing. Enamel colors would crack and flake off if the metal was not counterenameled. Transparent enamels fired over this coat of flux are clear and brilliant.

Another reason for using flux is that it allows you to draw the design right on the bare copper with black Underglaze D if you plan to do a large plaque. After you sift flux over the piece and fire it, the design will be clearly visible as long as you don't cover it with opaques. If you need no drawn design and you plan cloisonné work, set the wires right on the flux with Klyrfire and continue as explained when we discuss cloisonné (p. 54).

Flux for sterling: Sterling silver might need flux under certain parts, but not under blues, greens, grays, and whatever else your sample plaque will indicate. You do need *flux for sterling silver* under both opaque and transparent *reds.* Without the proper flux the opaque reds become black-spotted and transparents are completely discolored. Do not use a flux which is intended for fine silver (1000 silver, which is pure silver). It will turn yellow in firing and can be quite ugly after repeated firings. The firm of Schauer, in Vienna, has a flux available which is slightly blue, very clear, and can be used over *all* metals. This flux (#2A) tends to change reds a little toward blue.

Flux for 1000 silver: Fine silver presents the same small problem with flux under red, only here you need a special flux. Since there may be times when you need to fire flux over entire surfaces of either sterling or fine silver, it would be a good idea to have some of each flux on hand.

Flux for gold: Gold needs no flux if it is pure gold (fine gold) or 18 carat non-tarnishing gold. I recommend using no other gold for intricate work. It is possible to enamel on alloys (18 or 14 carat) if they consist only of fine gold, fine silver, and copper, but *no zinc.* You would then need a gold-flux, or you could use 2A. Make samples with this gold before applying enamel to the actual piece.

Flux under thin-rolled gold-sheet or gold-foil: If you wish to use a thin sheet of gold over the silver or portions of the silver, use flux. The same applies to copper: first fire the gold onto this thin and even coat of flux, then proceed as you would in enameling gold or silver.

Counterenamel is enamel fired on the back of the metal — it may be a plaque, the inside of the enameled base of a chalice, or the back of a piece of jewelry. On large plaques and pieces whose back will never be seen, you may use a mixture of leftover colors, but *not* what has been washed out of enamels. A good deal of the counterenamel may be flux. The counterenamel is sifted on the back of larger pieces, but counterenameling a golden pin or the inside of a ring should be done with the same care and skill as the front or outside. The reason for using counterenamel is to strengthen the front and to avoid cracks in the surface.

Counterenamel inside a ring.

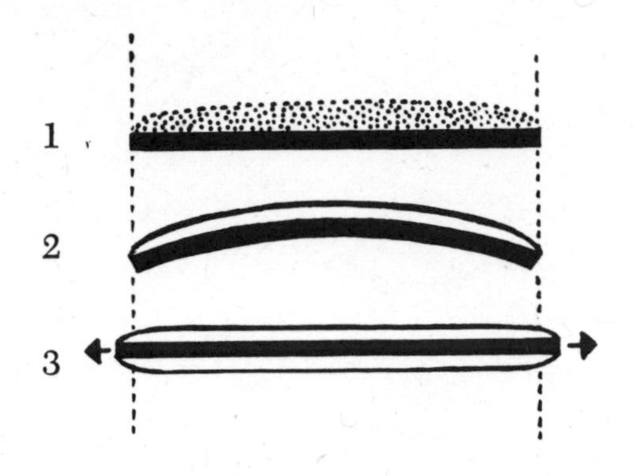

(1) Enamel sifted on metal. (2) When fired and cooling, the metal will warp. (3) If counterenameled, the metal stays flat, but "grows" sideways.

Metals, especially silver, expand and contract more than enamels when heated and cooled again. When the enamel is glowing hot, it is soft, like asphalt on a hot day. It covers the expanding surface of the metal. While cooling, the metal contracts much more than the enamel. The plaque will warp, raising the enameled side. This can come in handy if the craftsman expected it, but is very undesirable when the piece must fit and retain its shape.

If the back has a good coat of counterenamel, the equal pressure on both sides will prevent any warping. There is only one way the piece can expand: sideways. This is precisely what the plaque will do. *Enamels grow*, and this should be remembered especially when the piece has to fit a bezel. It is best to *finish the enamel first* and then make the setting. (See the note on "When To Counterenamel" in the Appendix.)

Opaque Enamels

Opaques let no light pass through. They could be compared to tempera or oil colors. It is good to purchase opaques ground to 80 mesh or even finer. They should be washed, but not as painstakingly and as often as transparent enamels. The appearance changes, the surface fires much better, as you will realize when you try both ways. Old, unwashed opaques can become quite disgusting and tend to blister when fired. Thorough washing and regrinding will restore them. Whenever one washes enamels it is quite upsetting to see how much goes down the drain. The waste should not even be used as counterenamel; it is no good. Compared to the small quantity of color actually used when working on precious metal, the waste is negligible and only the best and the cleanest is good enough. There are times when you want larger grains of some opaques, such as the red sparks in the panels on pages 92 and 93, and for this reason one should keep a few roughly ground lumps on hand and crush them with a hammer between strong paper. These are then "salted" into the wetpacked background for the desired effect.

Opal Enamels

Opal enamel is neither opaque nor transparent. It may be compared to the glaze on good china. Opals can be fired to an opalescent appearance. The first firing should be very high until transparent. Then the piece is cooled and fired for a second time at a lower temperature. It may be removed from the kiln and you can observe as the opalescent effect is achieved. It can then be fired again until you are content with the result.

Opal enamels need no flux except on fine silver.

The great charm of opals in contemporary work is their semi-transparency, especially the white opal fired over opaques to a marble-like effect.

For the opalescent effect I suggest that you buy the material in lumps. For overlaying and wetpacking, the factory-ground 80 mesh opal is good.

Transparent Enamels

Metal backgrounds and light are visible through transparent enamels. To keep them perfectly clear, transparents should be purchased in lumps and ground when needed. Only then will they fire to a pure brilliance. On large pieces of copper you can use preground transparents, provided they have been properly washed.

Translucent Enamels

Translucent enamel is even clearer than transparent. It is more like a slightly tinted glass and is used especially in *plique-à-jour*. Translucent enamels should be purchased in lumps and are treated like transparents.

MIXING ENAMELS

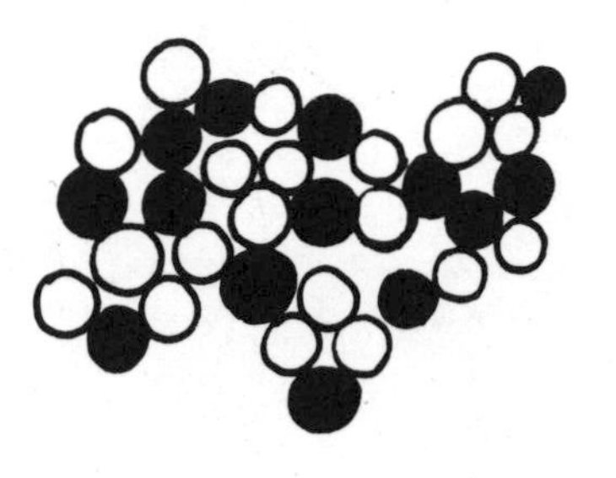

Enamels mix like pebbles.

Enamels cannot be mixed like paints, where black and white yields gray. Black pebbles and white pebbles, like enamels, will not combine to make gray pebbles. They will always remain small white and black specks when fired, but it may very well be that the artist wants such an effect. There is no limitation to the "textures" which can be achieved; textures which the eye can see but which cannot be felt since the stoned or polished surface will have the same smooth perfection in every part. After the first firing such mixtures may not look too promising, but shade them with transparents or cover them with clear flux or opal white, "salt" some opaque into the wetpacked transparents, and you will be amazed at how far even a limited palette can go with such techniques.

Good color combinations: The enamels you see in your jars or on the sample plaque are not yet "good color"; they are only the raw material. To get inspiration for fine combinations, watch nature: a single leaf in fall, moss partly covering a stone, clouds. It is all there, but it cannot be remembered accurately and be at hand when you need it in your shop. Make a few notes in your notebook just saying yellow, gray-yellow, brown-gray, pink-gray, etc., to bring the whole harmony back to you. I am not suggesting that you make pictures in enamel of the objects observed, but that you abstract their color and make a note of the appearance as transparent, opal, or opaque or whatever.

PREPARING TRANSPARENTS

I prefer to keep transparents and translucent colors in lumps because they will never deteriorate. Finely ground colors, even when neatly kept in jars, will change due to humidity. It becomes more time-consuming to restore these to good condition than to grind the necessary amount when you need it.

Put a small number of lumps into the mortar. A mortar made of china is adequate; agate is better for final grinding. Fill the mortar three-quarters full with water. The pestle is supposed to have a wooden handle, but I use a pestle made of china and have not had trouble with it yet. Hold the pestle barely above the lumps, then hammer with a wooden or paper-mallet vertically on the handle of the pestle, using short, sharp strokes. Keep the pestle moving across the lumps. You will be surprised at how soon the lumps turn into a coarse glass sand. When the water turns milky, pour it out and refill the mortar. Use the pestle to grind the enamel to about 50 or 60 mesh. You will soon learn how large you may leave the grains to fill them into the smallest spaces of your design and cover a surface evenly. Coarser ground enamel fires to a clearer transparency. While you are grinding, as the water becomes milky, you can easily pour this milky substance away since the heavier grains of enamel will settle to the bottom of the mortar. Continue to wash until the water remains clear. The same procedure with 80 mesh preground colors takes many more washings and will never yield the same pure, sparkling grains. Do the last washing with distilled water because tapwater contains too many chemicals which might cause trouble.

Washing enamels. Impurities wash out into the waste glass when water is poured into the spoon.

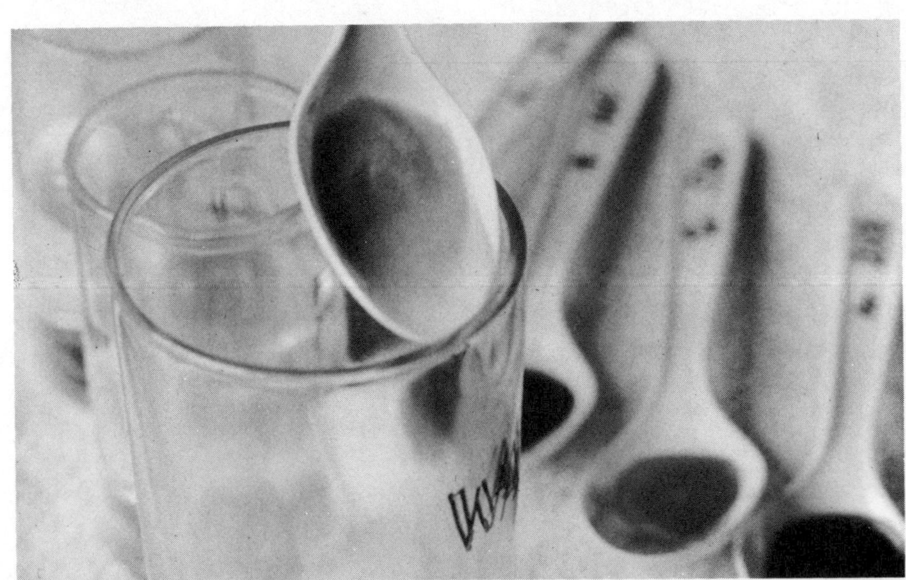

Crushing lumps in a mortar.

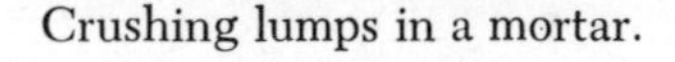

Grinding enamel under water.

GRISAILLE

Pegasus. Painted enamel in gold box.

Grisaille is an extremely finely ground white powder which can be applied with water, but rather is mixed with a very small amount of thick *oil of turpentine* on a piece of glass, using a horn spatula. Different oils are used as thinners, each affecting the reaction of the grisaille while working. Turpentine will make sharp contours possible while oil of lavender lets the grisaille flow. We can achieve effects similar to Japanese water-color painting with veil-like fine grays over the dark enamel surface. An extensive description of the technique is given on pages 71-74.

ENAMEL-PAINTING COLORS

I am far from suggesting china-painting on enamel, but there are times when it is good to be familiar with it. These enamel-painting colors are finely ground, they are prepared for work similar to grisaille and are dried and fired in the same manner as grisaille. If you are a painter, you will find some very exciting applications of these colors as sharp line designs between coats of transparents or over a fired enamel.

FINE GOLD AND FINE SILVER FIRED OVER ENAMEL

Fine gold (and fine silver) painted as finishing touches over grisaille enamel is very beautiful. This technique should be used with restraint. The gold is known to the refiner as "sponge" and to the china-painter as "Roman gold." It should be ground extremely fine between two pieces of rough glass, then mixed like grisaille with thick oil of turpentine. The application has to be quite thick but even. Fire until glowing dark red or barely glowing; then remove the piece and check to see if the gold adheres to the enamel. If so, polish the piece with a burnisher, or it can remain matte if treated with a glass brush under running water.

UNDERGLAZES

These are colors painted directly on the metal, dried, coated with sifted transparents, and fired. I use Underglaze D.

Apollo and Daphne. Grisaille box. (Owned by Mr. Chauncy Stillman)

DEGREES OF HARDNESS OF ENAMELS

Soft enamels melt at about 1300-1360 degrees F. (704-738°C.). They are used on metals which cannot stand high temperatures (sterling silver) or to fill indentations at final firing when the piece should not be exposed to great heat.

Medium enamels melt at about 1360-1420 degrees F. (738-771°C.). Most of our enamels are medium hard. If you use care and keep an eye on the work, they will be very satisfactory on sterling, fine silver, gold, and copper.

Hard enamels melt between 1420-1510 degrees F. (771-821°C.). Their melting point is close to that of silver. Therefore, they should be used only on high-carat gold or copper, or in work combining both.

Avoid firing hard enamels over softer ones. They will not fuse with the layers beneath and will chip and flake off.

FIRING TEMPERATURES OF ENAMELS

The temperature at which the enamel melts must be below the melting point of the base metal. Let us compare these:

Sterling silver	1640°F.	(893°C.)
18 carat gold	1700°F.	(927°C.)
14 carat gold, depending on alloy, *approximately*	1760°F.	(960°C.)
Fine silver	1762°F.	(962°C.)
Non-tarnishing 18 carat gold	1810°F.	(988°C.)
Fine gold	1950°F.	(1066°C.)
Copper	1981°F.	(1082°C.)

Temperatures inside the kiln:

Dark red	*approximately* 1300°F.	(704°C.)
Cherry-red	1400-1430°F.	(760°-780°C.)
Bright cherry-red	1450-1500°F.	(788°-816°C.)
Light red-orange	1550-1600°F.	(843°-871°C.)
Orange	*over* 1600°F.	(870°C.)

With these tables you will easily understand why sterling silver, for example, has to be enameled with medium or even soft-fusing colors. Gold and copper, on the other hand, are beyond danger as long as we don't forget them in the kiln – for then even gold may melt.

FIRING ENAMEL

When the decisive moment of firing enamel has come, preheat the kiln to a bright cherry-red (1450-1500°F.). Pyrometers have a tendency to vary. It is better to check with your eyes and not let the kiln overheat when you are working with silver or silver cloisonné. Perhaps there should be a sign over your kiln: DON'T TALK TO ME. There should be only one thought in your mind: setting the trivet and your piece very gently into the center of the kiln. It is best to set it deeper into the kiln rather than close to the door; the heat is more constant further inside. This must be done carefully — without losing your balance, dropping the piece, spilling enamel over the interior of the kiln, or jarring the little cloisonné wires which are held to the glassy surface only by glue that burns off the moment it enters the kiln. Complete concentration is most advisable.

If the kiln door opens downward, rest the rod of the firingfork carefully on the open door front, and then slowly let the fork part with the trivet and your work down inside the kiln. In this manner you avoid any shock. The procedure is fast and not too much heat is lost. All opening of the kiln should be done fast since the temperature drops at once 100 degrees or more.

There is no rule for firing times. However, if you have to repeat a piece and wish to eliminate the guesswork, take notes on the time to fire and the temperature which produced the best result.

While the plaque is still in the kiln, observe it and remove the piece when the enamel has fused to an even, shiny surface.

If the enamel is underfired, the surface still looks like "orange peel" and the transparents are dull. If cloisonné is underfired, the colors will creep to the middle of the cells, away from the rims, or cloisons. Tarnishing metals would get dark, but if you put the piece right back into the kiln and refire it, the enamel will smooth out. If fired at a rather high temperature, the enamel will seep up the sides of rims, or cloisons. If you overheat silver and silver cloisons, they disappear into the enamel, melt over the copper base, or turn into very ugly black crumbs. Such a piece cannot be saved. This has happened to many artists — and one can always repeat the work.

Never become too anxious when taking a piece out of the kiln. Concentrate on the task or you may invite bad luck! If your luck is bad — and this is another experience everybody has — the enamel slips from the trivet or falls outside the kiln on the floor. If it slipped from the trivet inside the kiln, take it easy. With asbestos gloves and a strong pair of long pliers or whatever tool is on hand, remove the piece from the kiln. Avoid letting any enamel touch the heating elements. Enamel covering this wire spiral is disastrous: the metal will break.

Enamel on the bottom of the kiln will get hot and sticky each time the kiln is in use. Trivets get stuck to it; it is truly unpleasant. Use a material called "kiln-wash," a white powder which repairs the damage when laid over the soiled spot. Though it is mixed with water, it doesn't really wash the kiln, but covers the soiled bottom.

Enamels which fall may be warped if they were hot. This may not be too serious if they are constructed as explained in the chapter on "Metal Shapes for Enameling." Reheating and a firm approach with asbestos gloves while the piece is still glowing can do wonders, but you may need another pair of helping hands.

Never remove a hot piece from the trivet forcibly. If it is stuck to the trivet, wait a little. If you tear it off, the hot enamel will be torn off too, forming long threads of glass from the piece and leaving bare spots.

By the way, never drop a hot enamel into water, no matter how impatient you may be, for the enamel will surely crack.

Left alone to cool, it will come off the trivet easily by itself and leave only tiny marks. If the metal was preshaped as suggested in the chapter on "Metal Shapes for Enameling," you will never run into this problem. There is always a clean metal edge on which the piece can rest.

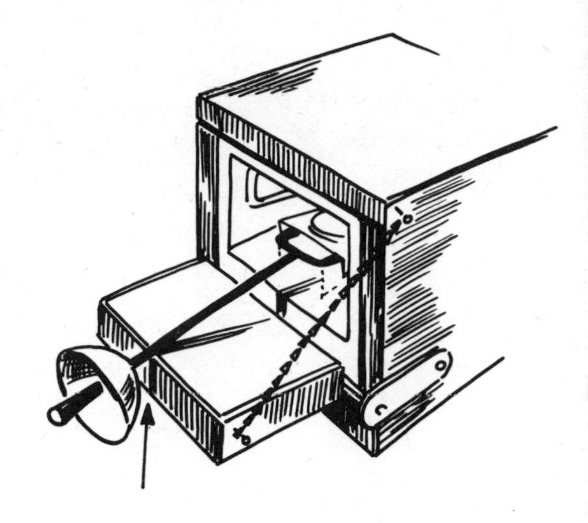

Rest the fork on the open door front and then carefully let it down inside the kiln.

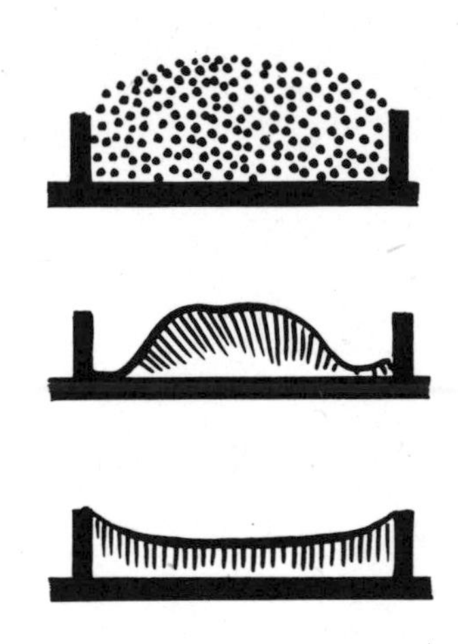

Well-packed enamel crawls to the middle of cloisonné cells when underfired. When fired fast and high, it seeps up to the rim, covering the metal (and cloisons) well.

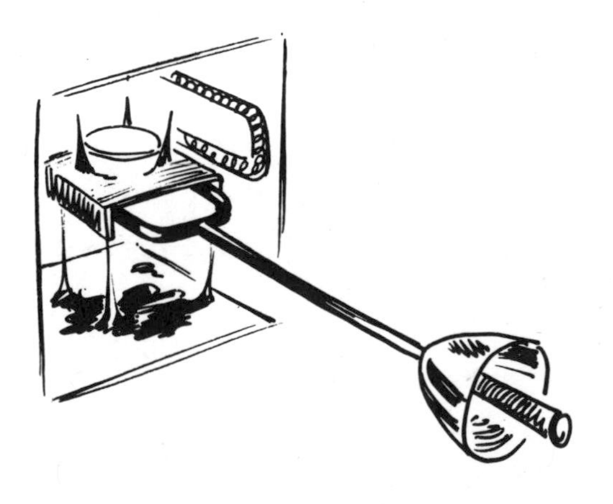

When the bottom of the kiln is dirty, the trivet may stick in the honey-like enamel.

Trivets of all shapes may be cut from stainless steel strips.

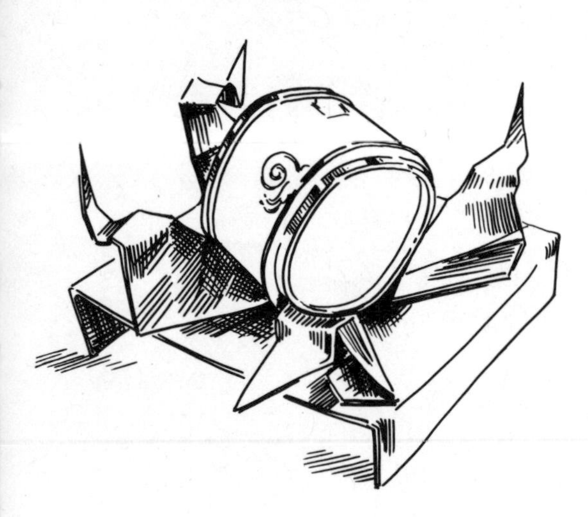

See that the enamel is not touched by the trivet.

Trivets

Trivets are ugly, dark-stained, distorted, mistreated, yet utterly useful and patient things. But their usefulness increases and, as the drawings show, even a band of stainless steel can be bent to serve innumerable purposes.

There are many good shapes on the market, but you will still have to invent your own or bend the available ones to suit your purpose. Even in the cleanest shop trivets become splattered with enamel, and this enamel sticks to your work. To clean trivets, file their edges sharp, hammer on them, or heat them glowing hot and drop them into cold water. They should always be ready for the next use.

It is most important that the enameled piece rest squarely on the trivet. It should have no chance to slide or lose its shape. Remember that all tension will leave the metal when heated; it will just droop down until it finds a hold. A hot hand-iron may be a useful tool if a plaque is warped or if the rim of a bowl is uneven. If the plaque is large, keep two sheets of asbestos at hand. Take the panel out of the kiln and place it on one of the asbestos sheets, covering it with the other (the asbestos will keep it hot for a little while). Then press with the hand-iron or some heavy object. If the distorted piece is a bowl, do not press too hard and keep it upside down on the asbestos.

If the metal is shaped as suggested, there is always a clean metal edge on which the piece can rest.

THE SAMPLE PLAQUE

For the experienced craftsman as well as the beginner, samples of all his colors are essential to show how these will appear on various backgrounds. Only by making such sample plaques will he find out what a large variety of colors just a few enamels can produce.

In a small workshop one panel of copper, 4 x 6½ inches, is enough for the color samples. In a larger shop you may prefer to have each color on a separate plaque.

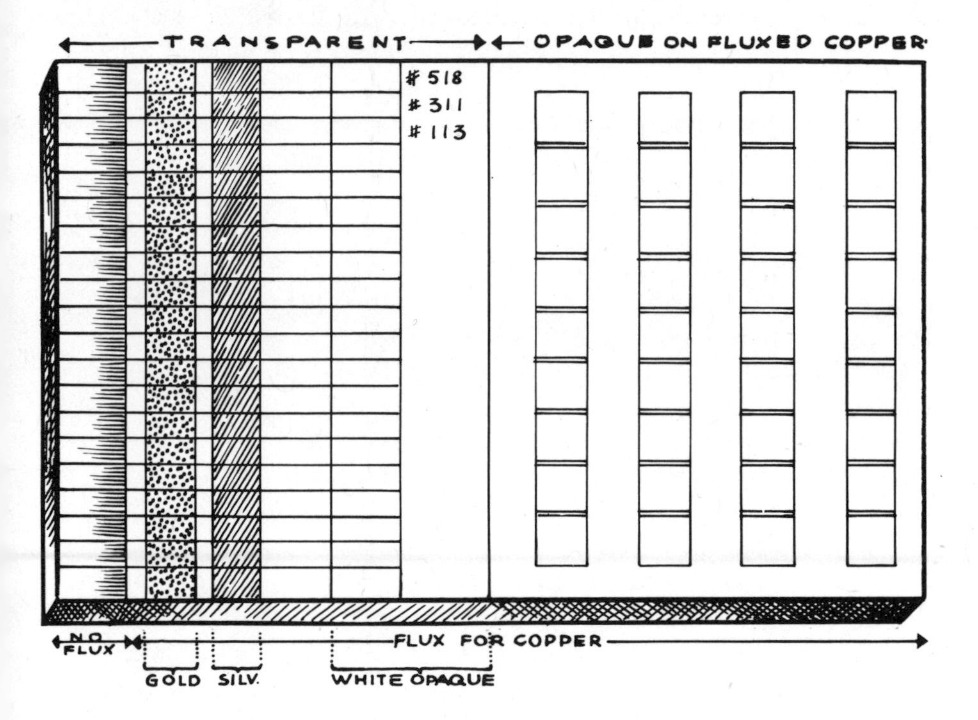

1. Start by shaping the plaque properly, edges bent down. Clean the surface in pickle (sulfuric acid bath: 1 part acid to 10 parts water). Then brush under running water with a glass brush, *never* with steel wool. Steel wool has no place in the enameling shop. It is greasy and leaves small particles of iron which contaminate enamels and work area.

2. Scratch strong lines vertically and horizontally into the plaque with a scriber. You need seven vertical lines on the side where the transparents are to be applied, and as many horizontals as you have enamels. These lines will remain visible under a coat of copperflux.

3. Paint the vertical stripe at the far left and the back of the plaque with Scalex. Scalex prevents the cinder and oxide, which appear on copper in high fire, from flaking off. If Scalex is not used, small particles of cinder will fly onto the hot enamel leaving black spots which are very hard to remove. If cinder gets into a jar with enamel, the entire color has to be discarded! It can never be separated again. Scalex should be used whenever bare copper is inserted into the kiln.

4. Cover the strip of Scalex in the front with paper and sift a coat of copperflux over the entire plaque. Use it generously around the edges, for here the enamel burns away first. Fire the piece fast and high until the flux looks golden.

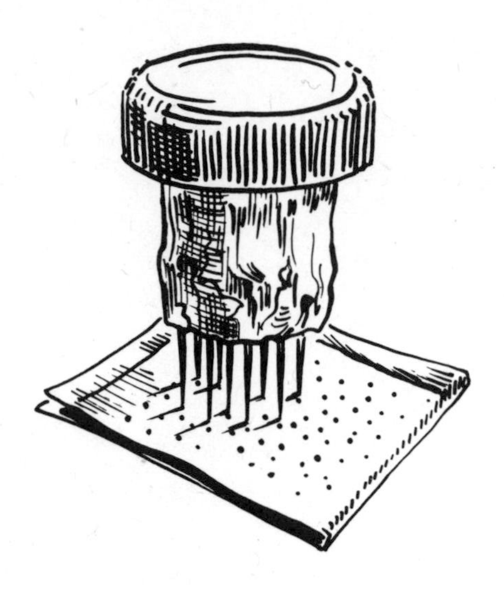

Perforating gold foil with a tool made from a cork into which fine sewing needles are inserted.

When the plaque cools, the Scalex and the oxide will peel off in one sheet. Be sure that all of this is removed carefully. The copper will be quite clean and for a background which will not be seen it is good enough.

5. Sift a medium thick layer of counterenamel over the back. But *do cover* the small strip of bare copper in front first with Scalex. Fire again, not necessarily to maturity. The flux on the front might burn away around the edges.

6. When cutting gold or silver foil, keep the foil between the thin sheets of transparent paper. While the foil is still protected inside this thin paper, punch small holes into it with a fine sewing needle. These tiny openings let any air which may have been trapped under the foil escape. A primitive but useful tool is a cork into which a number of such needles are placed, all sticking out the same length.

Paint Klyrfire over the two sections where gold and silver foil are to be used. Then lift the foil and the top layer of paper with a pair of tweezers and gently lay them into place. Foil is sensitive; it should not be pushed around. The paper will come off easily after you have dapped it (and the foil) tightly to the flux. Without the protection of this paper the foil would stick to your finger.

The bare strip of copper needs another coat of Scalex before we cover the plaque with paper. Leave only that portion of the piece exposed where you need to sift opaque white, then remove the paper.

7. Now place the plaque on a trivet or wire-mesh and fire it until the white has fused and the gold and silver adhere. The enamel must be dry before it is placed in the kiln.

Cleaning a strip of copper on the sample plaque by dipping it into pickle.

8. The next step is to clean the small strip of bare copper. Immerse it into Sparex or sulfuric acid bath (1 part acid to 10 parts water). *Always add acid to water, never water to acid,* or it will boil and develop poisonous fumes. Only the copper should be in the acid, the whites are sensitive to it. You can also clean the copper with emery paper or an electric tool with abrasive attachment. Rinse and wash the plaque thoroughly to get rid of all possible impurities.

9. The plaque is now ready to receive the transparent and opaque samples. Wash a small amount of each color and wetpack it with a small paintbrush in stripes which cover first the bare copper, then the flux, the gold, the silver, and half of the white. A spreader helps to distribute the enamel evenly. If it is too wet, soak out some of the moisture with Kleenex or some old clean linen, just held against the edge of the color. Push the enamel into its confining line which is showing through the flux. Apply a stripe of each transparent in this way.

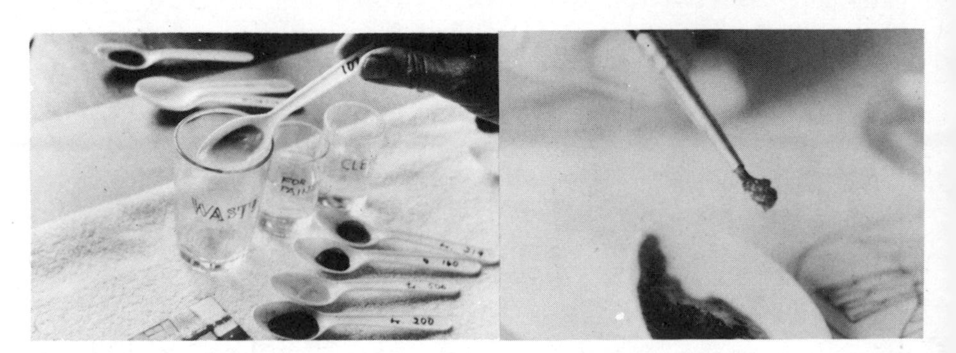

After the enamel is washed, it is wetpacked with a small paintbrush.

Let the colors touch each other but not run together. This is an excellent time to learn wetpacking. If the enamels are too dry, add a few drops of distilled water with an eyedropper. Later, write the number of each color and where it was purchased next to each sample in the remaining half of the white field. While applying the enamels to the plaque, please take notes of the succession of colors.

10. *Samples of opaques* need no gold or silver foil since they will be nearly the same shade no matter what the background. It suffices to apply them in small squares, stripes, or any manner convenient to the craftsman. The transparents and the opals will delight you with their variations. Some are beautiful directly on copper, some are good on silver, others are not. The sample plaque shows exactly what can be done and by which method to achieve it.

Many craftsmen fire small samples of each color and attach them individually to the bottle containing the enamel, or they invent a system which lets them choose the right color from single samples one inch square. I would miss the many variations of one color and the opportunity to select from the entire range of possibilities.

SPECIAL SAMPLE FOR IMPORTANT WORK

Even if the craftsman knows his colors by heart and knows what they can do, it is wise to make special samples for each new task to make sure that the colors will be in harmony with each other, that they have the same melting points, and that they will achieve the desired effect on the metal he intends to use.

SAMPLES OF SKIN SHADES

A 3 x 3 inch square of copper is large enough for a sample. Counter-enamel after the front has received a coat of flux, then wetpack vertical stripes of all the ivory, tan, beige, light brown opaques you have. Fire these to maturity.

Over these verticals wetpack stripes of all fine transparent grays, tan, light brown, brown-olive, etc., and fire again. When you realize that a final coat of clear flux will draw the harshness of colors together and blend them harmoniously, you have a nice palette for figure and portrait enameling.

SAMPLE FOR SKIN SHADES : ↓ ↓ ↓ ↓ ↓ ↓ ↓

		↓	↓	↓	↓	↓	↓	↓
TRANSPARENT	106 T.C.T. →							O
"	129 " →	"	"	"	"	"	"	PA
"	130 " →							QUE
"	131 " →	6	6	5	4	2	2	:
"	132 " →	2	9	9	0	5	7	WHI
"	179 " →	8	5	5	2	2	7	TE
"	309 " →	4	1					
"	85 SCH. →	SC.	SC.	T.C.	T.C.	T.C.	T.C.	
"	FLUX →	H.	H.	T.	T.	T.	T.	

To make samples for skin shades: Fire flux over copper plaque to golden brilliance. Wetpack the opaques vertically and fire. Then wetpack an even coat of the transparent, shading horizontally over the opaques; write with *jet black* the numbers and source of supply on the left part of the plaque, and fire. This sample gives you 63 different shades from which you may choose. (T.C.T. = Thomas C. Thompson; Sch. = Schauer, Vienna)

CAIN UBI
EST ABEL

REDS AND SOME TRICKY OPAQUES

All the reds demand our special attention. *Opaque reds on copper* are no problem. They fire perfectly even and are beautiful. But use the same red on silver and it will turn out black rims and black spots without flux. In cloisonné work the coat of flux or white under the red should be thin; otherwise you might stone through the red and end up with red-spotted flux or white enamel. Still, when using silver, it is hard to avoid the small black edge around each red area; more precisely, it cannot be avoided. At times, however, it can be rather attractive.

Transparent reds are the real problem. They should be ground coarser than other colors for best transparency. On copper, if fired only once, some are extremely beautiful. To prevent flux from firing through in small golden dots, mix one half flux and one half red for the first coat. Then add the second coat of red and the result will be good. If the piece has many colors, complete it except for the areas where the transparent red is to go. There leave just the flux over the copper. Add the reds before the last firing, which should be fast and high.

When you work with silver, use the proper flux for the alloy. Again, add the reds at the last firing. Reds become brownish when fired too often.

No problems exist with red transparents directly on gold. No flux is necessary. This means that you can get the perfect red on any metal if you underlay with gold, either a very thin-rolled gold sheet, which will not wrinkle, or foil, which will wrinkle.

Pink transparents are even worse than reds. At the first fire, some are as gentle as pink can be. Fire them again and they turn into a bluish dark pink. On copper they should be fired over opaque white and again only once or twice at most. Over gold, pinks are delightful, but change to orange-pink. Cover a pink with opal white as a last coat and only poetry can describe the result. Some sea shells have such hues on the inside.

White, of course, is opaque except for opal white. It is affected by acids and becomes dull. Soft white fired very high directly on copper without flux turns into very beautiful gold, green, and brown where the heat began to eat it away. This could be an inspiring background (for gold cloisonné) if very light transparents are used in a free and imaginative manner. Opaque white should be ground very finely for greater density. If a white-enameled piece turns yellow around the edges, it was overfired.

Black opaque is best fired only once. It too does not agree with acid. To get a spotlessly clean black is no easy task.

If opaques appear dull, they have either been in a kiln at too low a temperature for too long, or they have been exposed to acid or pickle, or the enamel was old and not washed.

Red is fired last over flux.

Care must be taken not to stone off the top layer of enamel.

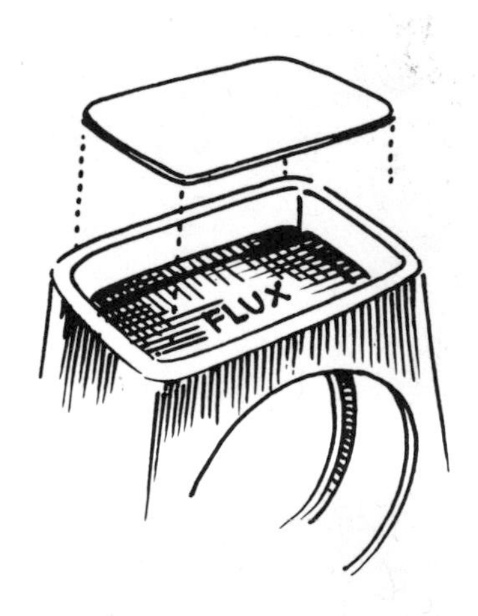

A thin sheet of pure gold is fired over a coat of flux; this ensures a brilliant red.

Cain and Abel. Detail of plaque in silver cloisonné. Height: 4 inches.

3.

Metals for Enameling

COPPER

Copper is the best carrier of enamel for larger pieces or for those which need a sturdy construction. Copper and gold cloisonné is a good combination since neither expands as much as silver. Copper takes gold-plating very well. Copper *does* oxidize heavily and bare areas (without enamel) must be protected with Scalex. Copper cannot be cast. When using copper cloisonné wire, the piece must be brushed with a glass brush under running water after each firing to remove the cinder. When all wires are coated with enamel, this cleaning is no longer necessary.

TOMBAC

Tombac looks like copper. It can be cast very well, and also can be enameled without problems if it is not fired too often. Tombac is an alloy of 95 percent copper and 5 percent zinc. Since it can be cast, this metal offers great possibilities to the sculptor-craftsman.

Pewter crucifix with copper cloisonné stoned to a matte finish. Height: 12 inches. (Owned by Hans Zeitner)

STERLING SILVER

Sterling silver can be enameled easily, and it can be cast. Sterling casts have a white coating of fine silver which should be kept intact except for the area which is to be enameled in the next firing. Such casts do not tarnish except where the coating is damaged. Sterling silver does tarnish otherwise. The place to be enameled reflects better if subtly engraved or Florentine-finished.

FINE SILVER

Fine silver (1000/silver) is pure silver. It does not tarnish. It expands more than copper or gold when enameled, is almost ideal for enameling, and brings out the brilliance of transparents (excepts reds) without flux. It is rather soft and should be used for enameled parts only. These parts can then be set into more resistant metal like precious stones. It is precious enough to be worth the investment of time and effort. Fine silver cannot be cast.

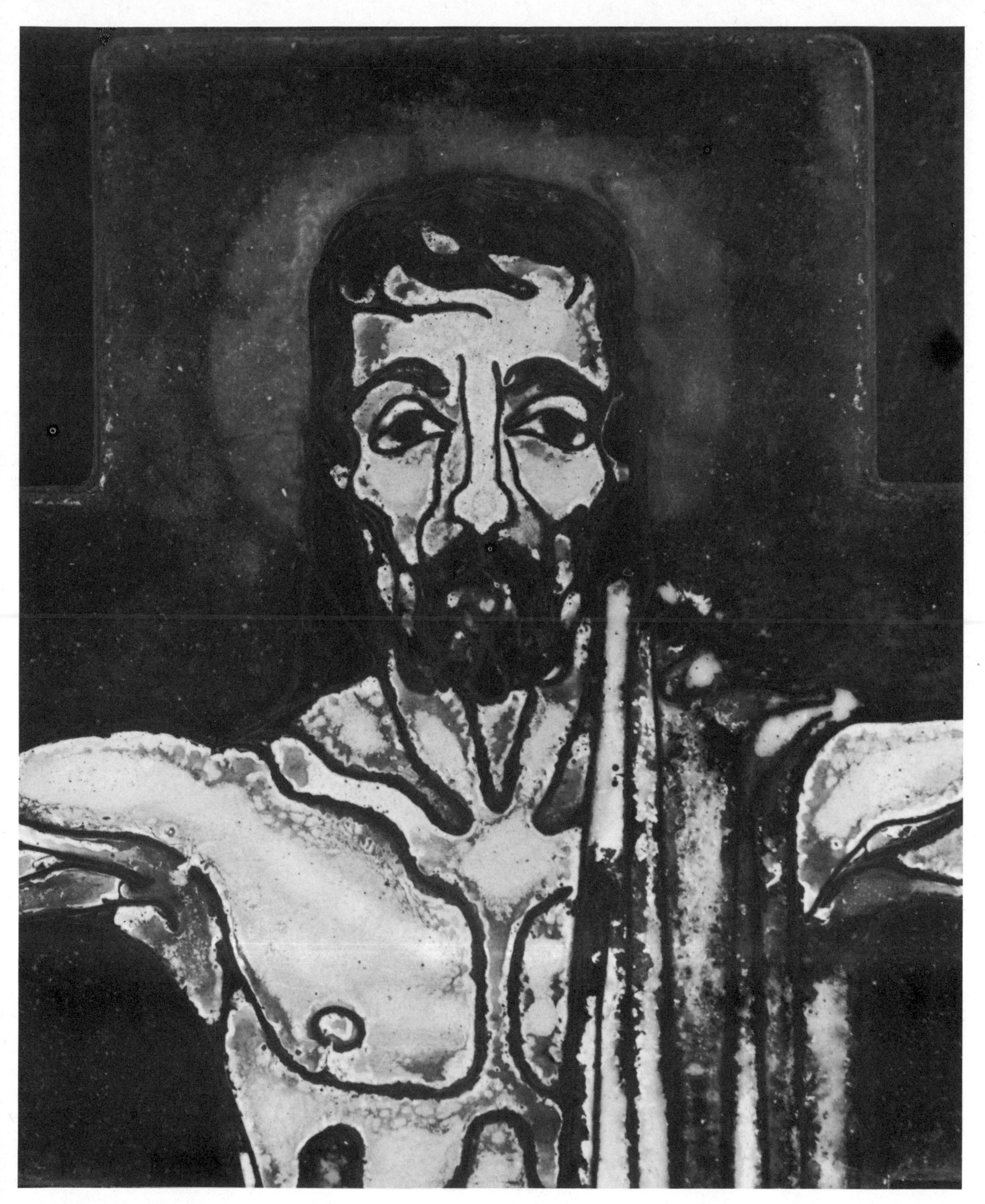

Detail of crucifix shown on facing page.

EIGHTEEN CARAT GOLD

Eighteen carat non-tarnishing gold (N-T gold) is an alloy of 750 parts fine gold and 250 parts fine silver, no zinc. Both metals are ideal for enameling. This greenish gold is rather soft and should be used like fine silver, as carrier for enamel rather than for objects which are exposed to great wear and tear. It is difficult to cast this type of gold, but it is possible. It is very suitable for cloisonné wire and can stand high temperatures.

FINE GOLD

Fine gold is pure gold (24 carat). It is *the* metal for exquisite enameling. The color is different from that of 18 carat non-tarnishing gold. Fine gold is more yellow while N-T gold is greenish.

Cloisonné wire can be made of all these metals except for sterling silver which melts easily and does tarnish.

SUGGESTIONS FOR THE THICKNESS OF METAL TO BE ENAMELED

Copper

Large copper panels (over 12 x 12″)	14-16 ga.
Sizes up to 12 x 12″	18-20 ga.
Small shapes (up to 1½ x 1½″)	20 ga.

Very thin copper does not keep its shape well.

Silver

For the carrier metal	20-22 ga.
For the top piece, which is sawed out (champlevé)	18-20 ga.
For rims of square wire to be soldered around the edge of the carrier metal	16-18 ga.

Gold

Similar to silver, it should be high enough for cloisons. See silver.

Copper cloisonné wire can have any thickness, depending on the size and character of the piece to be executed. It should always be at least 1 mm high.

Silver cloisonné wire should be about 1 mm high, 0.15 to 0.25 mm thick.

Gold cloisonné wire should be about 0.8 mm high and 0.10 mm thick.

HOW TO MAKE CLOISONNÉ WIRE

There is no problem in buying cloisonné wire in silver, but it seems quite difficult to find the non-tarnishing gold, and fine gold cloisonné wire is very expensive in small quantities.

Let me give you a few suggestions on how to make your own wire:

Take any piece of the proper gold (fine or 18 carat N-T). The piece should have no sharp corners and be about 18 to 20 ga. thick. Saw in a spiral, parallel to the outside edge forming a band of metal as wide as it is thick. Use a fine sawblade to avoid loss of gold. Saw until you have a long enough piece or have reached the center. Anneal this long band of gold and pull it through a round or square draw-plate until the wire is about $\frac{1}{3}$ mm thick. A round draw-plate is preferable. If you (or your school or a colleague) have a rolling mill, you need only roll this thin round wire to 0.10 or 0.15 mm. The third of a millimeter will be just right to get a wire of 0.8 mm height. If there is no rolling mill available, take a good planishing hammer and a small, well-polished, flat anvil and hammer with light and fast strokes, watching carefully that you always strike the same area of the anvil while your left hand moves the wire along under the hammer. This way the strokes will overlap closely and you get a flat wire which stands well as cloisons, although the edge may not be quite as straight as that of wire made in the rolling mill. Anneal before starting to bend it.

Sawing gold in a spiral to make cloisonné wire.

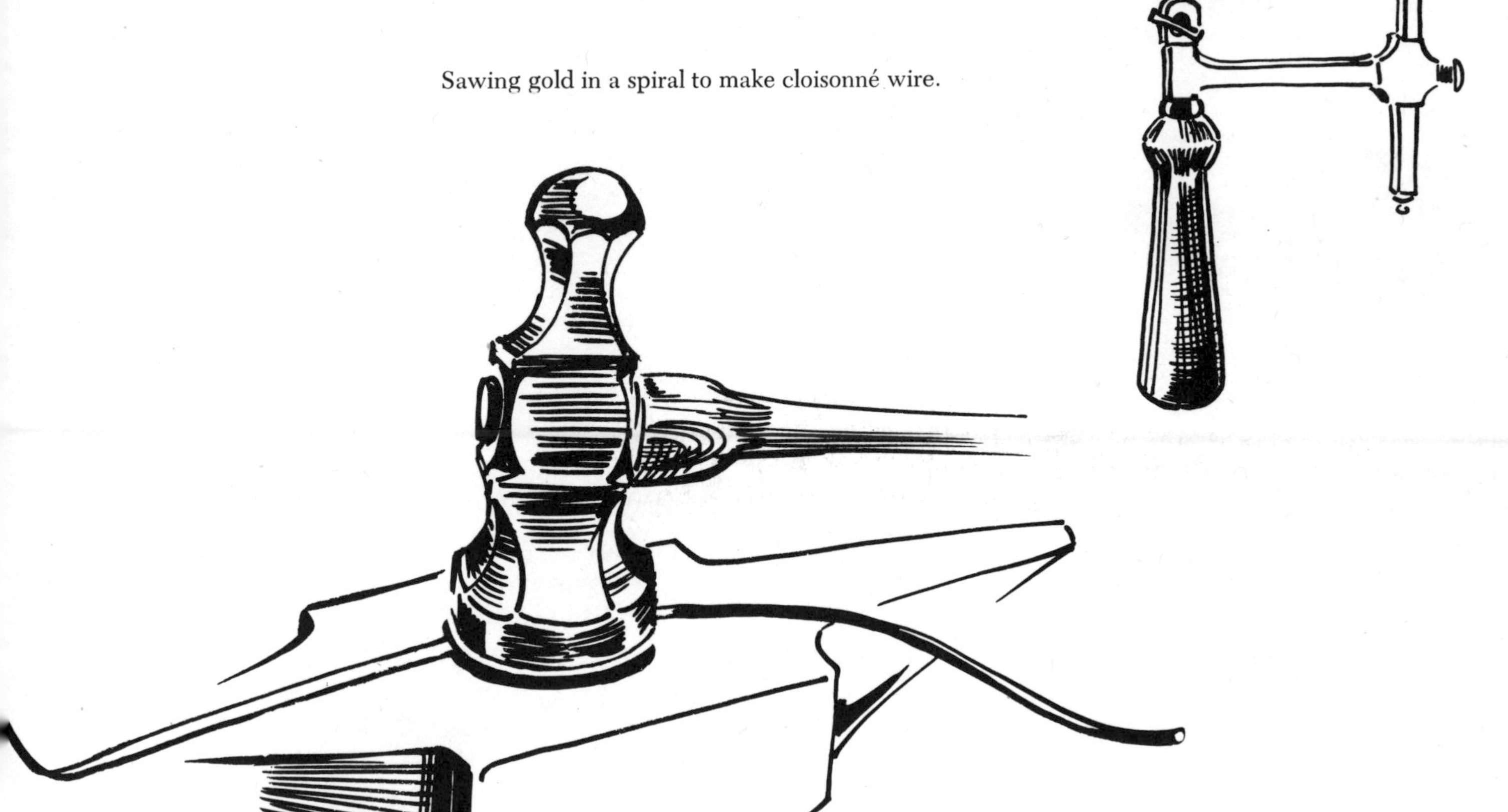

Hammering cloisonné wire.

ENAMELING 14 CARAT GOLD AND ITS PROBLEMS

If the 14 carat gold consists only of pure gold, pure silver, and copper, it can be enameled. The gold will tarnish and you work as you would with copper. You need flux. Transparents will not be as good as on high-carat gold. It would be better to work with opaques and, if there is to be some brilliant transparent, it can be underlaid with fine gold, sheet or foil. If you design the piece so that the enamel is done on a non-tarnishing metal and then set into the piece of 14 carat jewelry, you will have fewer problems and the result will reward your efforts.

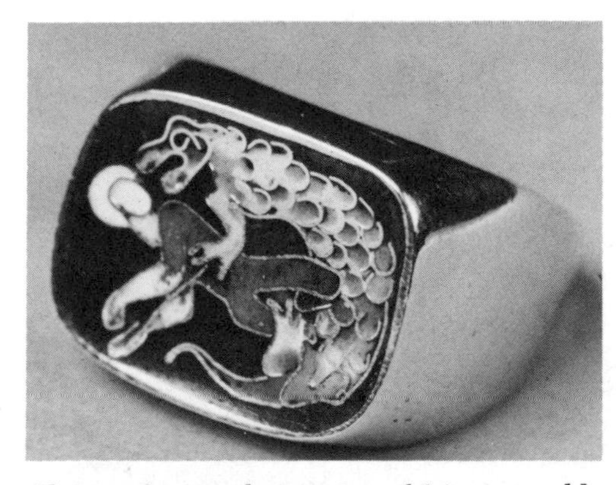

Cloisonné enamel on a ring of 14 carat gold.

4.

Metal Shapes for Enameling

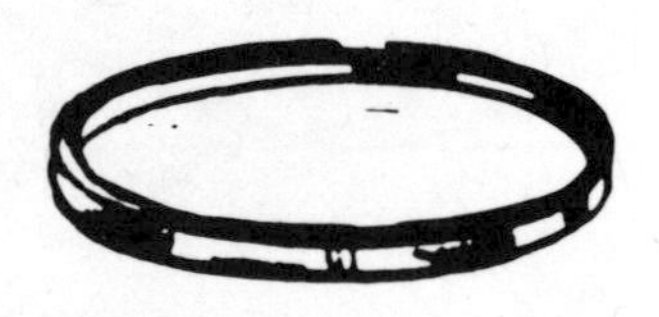

Making a rim for a plaque by tapping it down over the flat face of a hammer.

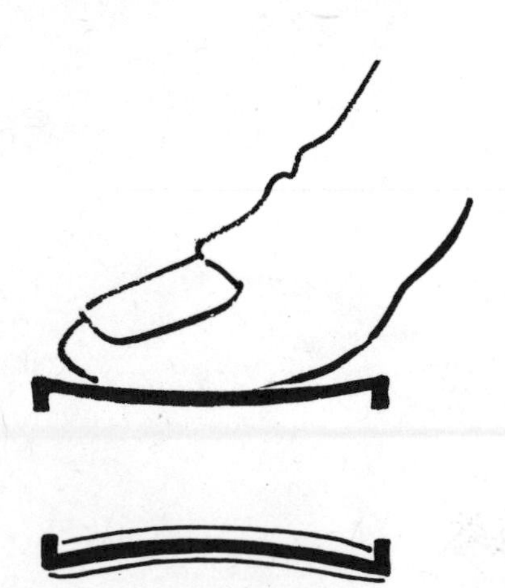

For a small plaque, thumb pressure may be enough to make the surface slightly concave, which is best for enameling on both sides.

FLAT SURFACES AND SHAPES

It is possible to wetpack or sift enamel on a slightly domed piece of metal and fire it in a kiln or even over a blowtorch. If the enamel is thin and the metal thick enough, it will hold for a while. When one tries to set such a piece (not glue it) into metal bezel, it may crack sooner or later. Any pressure or blow against the edge and it certainly will crack or chip.

Let us figure out how the metal shape should be planned and constructed so that the enamel will resist years of hard use. Enamels should have a rim – and in most cases should be slightly domed. The rim is for protection, the dome for keeping shape in the heat of the kiln, for better reflection of light, and because every place can be easily reached when stoning if it is cloisonné enamel.

The simplest way to make such a rim around a plaque is to saw the base about $1/16$ of an inch larger than the enamel area. Draw a parallel line around the edge with a compass, about $1/32$ of an inch from the rim. Then hold it over a stake or the flat face of a hammer and with either a mallet or a planishing hammer turn this $1/32$ of an inch down. Actually that side will be "up." Anneal the piece again, then gently hammer or press from the center to the outside of the piece while it is resting on a flat surface until the center touches the support. If the plaque is small, you can do this with your thumb, a mallet, or even a planishing hammer if you use them gently. If you work on a larger piece, planish over a flat or slightly domed stake. Start in the center with harder strokes, then work around the center to the outside with decreasing force, until at the rim you barely touch the plaque with the hammer. The concave of this metal shape holds the counterenamel.

Somebody now might say, "Those precious antique gold-cloisonné panels from Byzantium had no rims." Oh yes they did! A wide rim of bare gold all around the deep-set area into which the cloisons were placed.

Our turned-up rim is not yet the solution for precious enamels unless they are mounted from the rear into some kind of frame. The next possibility is to shape a rectangular wire (16 or 18 ga.) to the desired outline, join the two ends with hard solder, and file the inner edge to a slant. This edge, which is to touch the metal of the base, should be filed to a slant in order to prevent the solder from spreading over the base-metal. Solder causes black spots, pores, and other mishaps under transparents. Fit the rim to the base and place the solder around the outside after you have clamped the two pieces securely together. When the solder melts, it will stay underneath the rim and not interfere with the enamel.

For soldering, place the work on some wire-mesh and heat from underneath while slowly turning the piece. The heavier metal and the lighter rim will be equally hot and the solder will run smoothly. Pickle the plaque, saw and file the edge, clean it with a glass brush, and you have a piece which is strong, holds its shape, protects the enamel, and can be set into a bezel. Filing the rim makes setting easier yet. Metal touches only metal; no pressure can damage your enamel.

There is still one visible solder-joint on the rim though. It might disappear under the bezel, but if you do not plan to mount the plaque in a bezel and this is to be the finished piece – let us think the problem over once more.

If this rim were cut out of a piece of 18 gauge sheet-metal, there would be no solder-joint. Such a method would also offer variations in design and represents the first step toward *champlevé*. (See p. 65.) Solder this rim and finish in the same manner as suggested with the wire rim. And let us take the next step toward perfection: solder a rim of wire (it can be thinner, round stock) around the back of the plaque. Now the counterenamel has its neat place and can be executed with the same care as the enamel on the front. If the piece is to be a pendant, leave a small lip of the sheet-metal at the top. You can later drill a hole there and put a ring for the chain through it. This saves another solder-joint. If the plaque is to become a pin, now is the time to add two small "islands" to the back: one for the hinge, the other for the hook or catch. (See the note on "Pin Stems" in the Appendix.) Solder these islands to the back at the same time as the edge. Use very little solder. The hinge for the pin should first be soldered to the one island, then protected with ochre, and soldered to the future enamel. Keep the islands free from enamel. It is permissible to add the catch with pewter-solder after the enameling is finished and the piece is polished. For truly exquisite work I suggest setting the plâque into a gold bezel which has all mechanical parts properly attached to the back with gold-solder.

The metal shape you have prepared is worthy of any precious enameling technique.

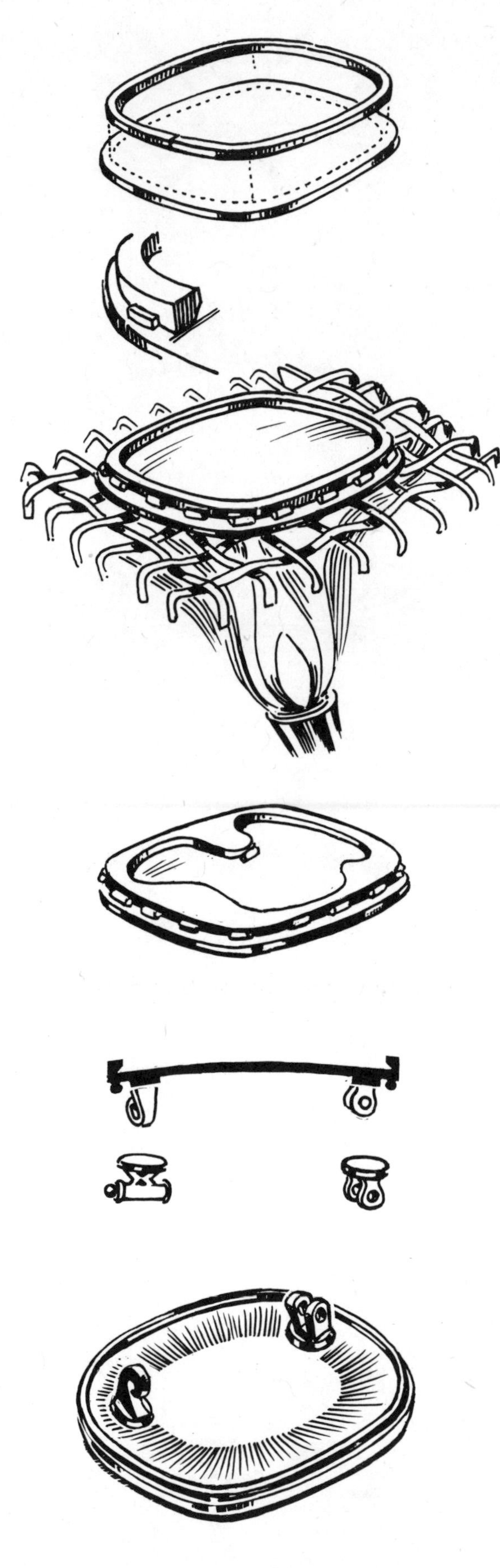

Preparing a small plaque with a rim to make a pin.

Enamel will crack if the bezel is not on a small metal island, as in the top example.

Setting a pearl above an enamel surface.

Setting a pearl deep in an enamel surface.

Setting a stone and pearls into an enamel surface.

Setting a large stone into an enamel surface. The outer vertical rim may also be enameled.

SETTING PRECIOUS STONES AND PEARLS INTO AN ENAMELED SURFACE

If you have prepared a piece as suggested and you solder one or more bezels for precious stones to its surface, the enameling will work well — until you try to stone the enamel, or try to set stones on it.

While stoning you would damage the bezels and when you set the stones, the enamel would crack under the pressure of the tool. The answer is to solder the bezels first on those islands which are as high as the cloison wire or as the thickness of the coat of enamel. Then solder these islands and their bezels onto the metal base of the enamel. There ought to be at least $1/32$ of an inch clearance around the bezel. Of course, this metal edge can be part of the design and can have any shape. Now you can enamel, stone, polish, and set the stone safely.

The sketches show how to proceed with the preparation of prong settings and pearls in different heights above the enamel. Think of pearls floating over a deep transparent enamel. There is no limit to imagination and invention.

It is the same task with variations: *protect the enamel and see that you can still apply counterenamel.*

MOVING LINKS WITH ENAMEL

1. The enamel, which has one edge turned up and a square wire rim soldered to the *back*, is the background to fine metalwork. It fits precisely when mounted from the rear, and is held by small metal rings which connect the links. Each of these small metal rings *must* be secured with a bit of pewter-solder when the links are assembled.

2. Another method for connecting links which are enameled and counterenameled is as follows: solder strips of flat metal (silver or gold of any carat) to the back of each link. These strips must be like two continuous bands when the links are placed next to each other. These two bands are part of the rim of the counterenamel. Two more small bridges have to be added below the two bands to prevent the counterenamel from chipping. Drill holes as shown in the drawing, large enough for 16 gauge oval or round wire rings.

When assembling the finished links, they should move well without interrupting the ribbon effect of the front.

3. Here is one other way to connect enameled links: include provisions for one or more openings in the design of the metal parts to remain in the metal base of the enamel, thus avoiding the need for soldering. When you assemble the links, connect them with flattened rings, ovals, or wide metal strips. These must be soldered with pewter-solder for stability.

4. You can also add hinges to each link while the metal part is being constructed. Protect the tubing of the hinge with ochre while enameling the piece.

1
2
3
4

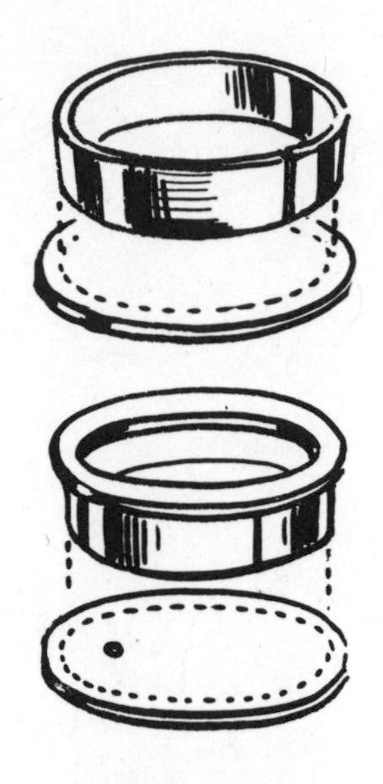

Making a band ring.

BAND RINGS

Band rings are the most rewarding exercise for the enameler who wants to get away from flat surfaces. If the ring is constructed in the manner suggested, the problem of solder-joints is nearly eliminated, as well as the possibility that these solder-joints might open and cause trouble.

Bend the vertical part, a ribbon of silver or gold, 20-22 gauge thick. It must be about one size larger than the desired size. Solder the joint extremely clean, using a hard silver or gold solder. Shape this band over the ring mandrel; flatten both sides and file the rims so that they slant slightly on the outside. Solder the band to a flat piece of 18 gauge metal; put the solder paillons around the outside. The surplus can be filed off easily. Saw and file the edge of this box-like piece, leaving about $\frac{1}{32}$ of an inch for the protection of the enamel. Then draw a line parallel to the edge with a compass, about $\frac{3}{32}$ of an inch from the rim. Drill a small hole on this line for the sawblade and cut out the center. Repeat this procedure on the other side of the band, then file both inside edges simultaneously, but not yet to perfection. This is where we will later fit the ring to its proper size, *before* the last firing of enamel.

The ring is then to be pickled and washed in ammonia to neutralize the acid, rinsed, and brushed with the glass brush. The outer surface can now be engraved, Florentine-finished, or treated with a ball-burr if you have a flexible shaft. The surface reflects transparents better and the roughness gives more holding power to the enamel. Your band ring is now ready for enameling.

If it is made of precious metal, you should stamp the carat *now*.

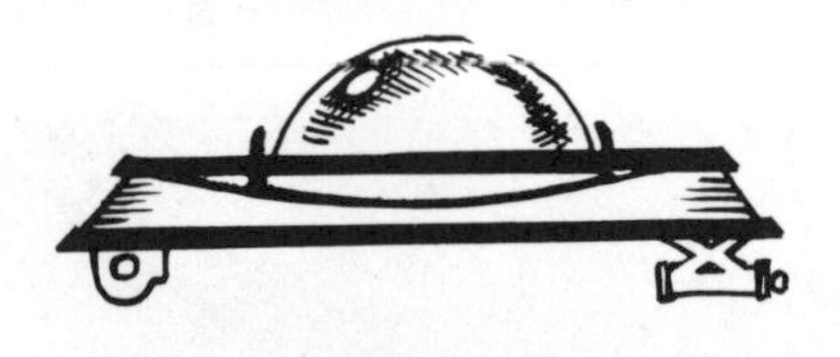

A SLANTED RIM

From a band ring to an enameled conical shape is a small step. If the piece is rather small, you can widen one side of it by hammering it over a ring-mandrel or a stake. If it is a larger object, it is better to draw a conical development and saw it out. Soldering of the band is again done with hard solder. Attach the bottom rim first, possibly leaving provisions for the findings when you saw out the center. Then add the top, which might be plain, or richly ornamented with metal, or even prepared for enameling with its own rim and with settings for precious stones – whatever you plan and already know how to do well.

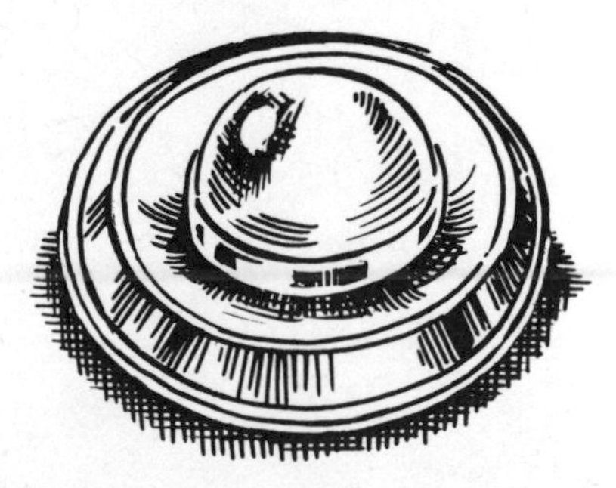

If you understand how to set stones and how to construct the metal parts of a ring, it is easy to make a brooch which has a rim enameled inside and outside.

A BOX, ENAMELED INSIDE AND OUTSIDE

Preparing the metal for a box, enameled inside and outside, is a small task if you understood how to make the two preceding pieces. But be aware of the fact that box covers must fit exactly and that enamels "grow" in the fire. Besides, the bottom needs counterenamel which should be executed as nicely as the rest of the piece.

1. Start again with a strip of metal for the sides of the box. The seam has to be very neat; it will show a little, but this is all right under transparents. Clean and buff the seam while you can reach it so easily.
2. Solder on the sheet of metal which will become the wide upper rim of the box; proceed as with the band ring. Leave enough rim inside the box, on the outside 1/32 of an inch is enough. This inside rim will be very important when you fit the lid of the box. File it to a good shape while the box has no bottom and you can reach the rim. Provide for the inside enamel and enough protection so that the cover will touch only metal.
3. Prepare the bottom, again leaving 1/32 of an inch all around to protect the enamel. *Do not solder it to the box yet!*
4. Prepare the metal for the enamel to fit the cover of the box; enamel and fire it. When it is finished you know the size for the bezel.
5. Make the metal part for the box cover with the bezel for the enamel. Prepare a strip of 18 gauge metal, 1/4 inch wide, which will fit into the rim of the box cover and will later be soldered to the lid. Fit this rim first; it must still be rather tight but ought to fit both ways. Now is the time to adjust both pieces; when you are satisfied with the fit, hold the cover over this rim *while it is in the box.* Scribe a line around the underside of the cover, add a witness-mark to be sure, then solder the lower rim to the cover of the box.
6. When this is done, you can enjoy finishing the box. Solder the bottom to the piece, add on its underside a wire or a flat band of metal (which takes the place of a foot and literally gives the box quite a lift) to hold the counterenamel; pickle, wash, and enamel.

May I add from my own experiences that it is a nice surprise if the inside, the bottom, and even the seam on the inside are finished with great care. Whoever touches such a piece cannot help but love it.

Preparing the metal for a box to be enameled inside and outside.

The box, ready to be enameled.

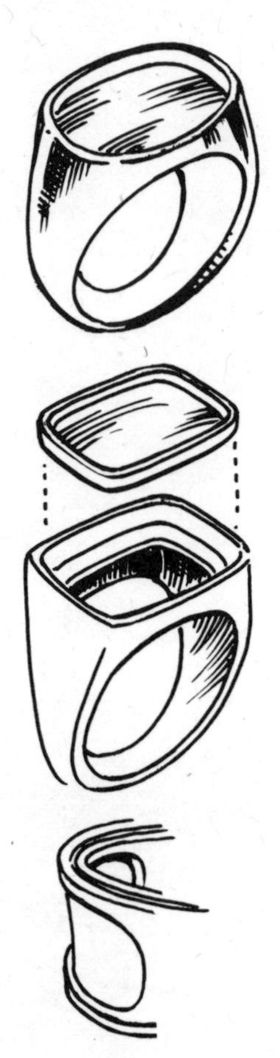

Top to bottom: a cast ring with platform for enamel; ring with opening for an enamel; ring with three layers of metal.

FOUR MORE OUTLINES FOR RINGS

1. A cast ring has a large platform with rather high edges ($\frac{2}{32}$ of an inch). This shape offers no problem when cast in sterling. There is space for counterenamel. The enamel on the face should be stoned slightly concave to give the enamel more protection. (See illustration on page 33.)

2. If such a shape had an opening instead of the platform on which to enamel, and if it provided a step inside on which to rest a separately made enamel on very good gold, this could be set like a precious stone. Also the ring can be made of an alloy of gold and yet contain an enamel on non-tarnishing gold.

3. The next ring is made of three layers of metal; the middle layer provides the base for enameling and should be non-tarnishing gold or fine silver. It doesn't make sense to apply so much work and skill and save on gold. (See *Adam and Eve ring*, page 101.)

The inner layer has a cutout similar to the outer layer for the counterenamel. When soldering the three layers of metal, leave enough width in the middle to hold the solder. File after soldering, but leave the finishing until the enamel has been fired.

4. A very decorative ring to be worn only on special occasions; for example, a bishop's ring:

First, make an exact pattern of copper or thin pewter.

Then saw it out of precious metal and solder with hard solder as explained by the sketch.

The shank of the ring may be domed slightly over a dapping-punch. File it smooth and solder on a flat piece of metal for the top. If you use gold, it should be non-tarnishing. File the two open sides of the shank to a fine curve and fit flat pieces of 18 gauge metal (it need not be non-tarnishing) exactly to these curves.

It is advisable to make a correct model of the desired shape. Discarded X-ray films are ideal for patterns and models; they are transparent enough to permit you to see the fit, yet elastic enough to bend. Ask your doctor for some.

Solder one side of the ring at a time, then saw out the center, leaving enough rim on both inside and outside to protect the enamel. Cover all soldered areas with ochre and solder a rim around the top piece, or a flat piece of 18 gauge metal for champlevé — whatever the design requires. It is a good idea to treat the surface which will be enameled with a Florentine-finish tool. This increases the reflection and gives a good hold to the enamel.

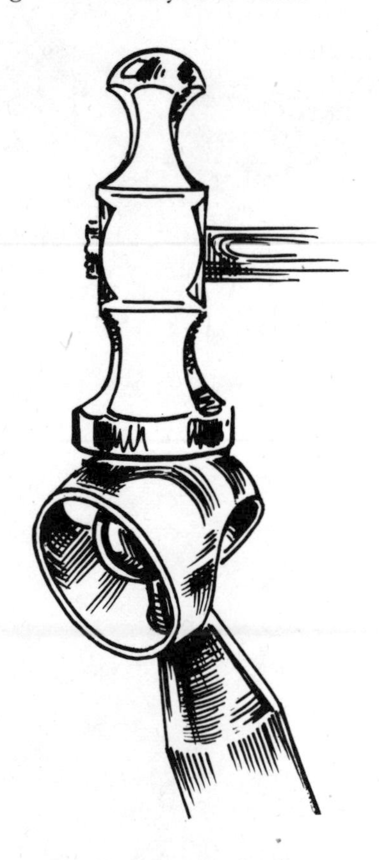

Doming the shank.

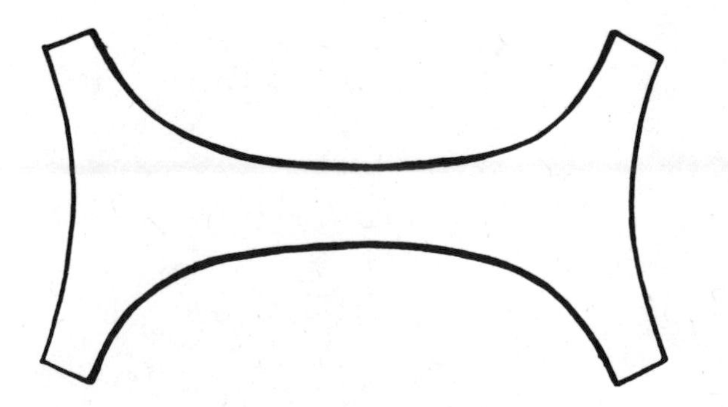

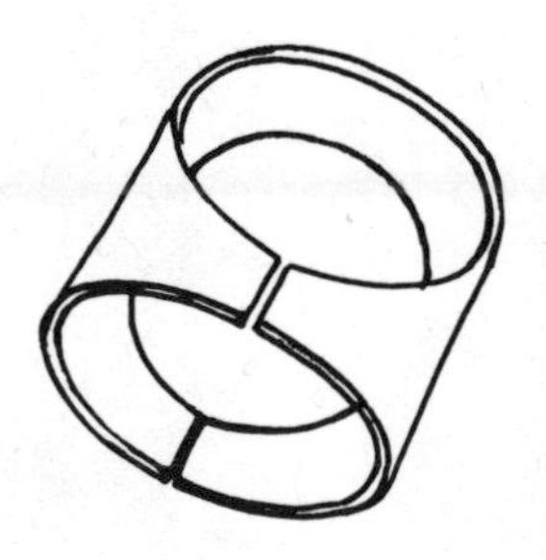

Pattern for decorative ring shown on opposite page.

The metal part for the bishop's ring, ready to be enameled.

TO FIT A SMALL ENAMEL INTO A CURVED SURFACE

When planning as intricate an object as an enameled chalice, it is wise to separate the metalwork from the enameling. Should the enamel ever be damaged, or should the cup need repair or new gold-plating, it is possible to separate the two without heating or destruction of any part. No soft-soldering, a few prongs to bend open, a nut to be unscrewed — should make any repair feasible. The three enameled coats of arms in the gold cup at right had to be mounted inside the foot of the cup flush with the outside surface.

1. To begin with, all metalwork had been finished as explained with the reproduction on page 42.

2. I made an exact pattern of the curve of the foot from pewter, which is the easiest to shape, using about one half of the foot. I mixed water and *asbestos stove-lining* and made a shape large enough to hold the enamels (one at a time), by pressing this paste into the model of the foot. It was much too wet to remove without destroying the fit, but I could soak a remarkable amount of water out of it by pressing Kleenex over the paste inside the pewter shape. When the consistency permitted, I gently pushed the asbestos form to a paper towel and placed it on top of the kiln for a slow drying.

The next day I heated it very slowly to let any moisture that was left escape. Then I fired this shape until it was glowing red. I put some small pieces of steel wire into it carefully (the substance is brittle) where the enamel would rest during firings. I was able to use this little piece during the firing of all three enamels.

Cup with three coats of arms enameled separately and set into the foot from the inside.

A base to keep a curved enamel in shape.

Clockwise from top, left: The plaque fits the curve exactly; an inside view; the asbestos model which serves to keep the shape during all firings of the plaque; cross-section; the plaque set in place.

3. The pewter model could now serve for the fitting of the three metal parts which were to be enameled.

I sawed out the "window" behind which the enamels had to fit. Then I shaped three pieces of 18 carat N-T gold, 20 ga. thick, to fit the "window" from the inside. Each of these curved pieces overlapped the "window" by ¼ inch all around. Four witness-marks on each piece showed exactly where they would be held by four prongs, which had to be soldered to the inside of the foot.

The opening of the "window" was marked with a fine line in front of each piece. Then I shaped and soldered a rim of 18 ga. square gold wire on each piece, parallel to the "window" line, but leaving $\frac{1}{32}$ of an inch distance to this line. Inside this rim is the enamel. The small distance, backed by gold, is attractive and makes a clean setting possible, since a perfect fit without any gaps cannot be achieved. Remember that every enamel will change its shape just a little.

This technique of holding an enamel can be applied to many other objects. When I speak of my own experiences, I hope that they will make the start easier for my fellow craftsmen.

A CONE OR CYLINDER TO BE PREPARED FOR CLOISONNÉ ENAMEL

Based on your design of the object to be made, draw a conical development on paper, sketch the design and color into it, cut it out and fold it so that you get an impression of how it will look in three dimensions. It will probably look much bigger. I strongly recommend that the design include some partitions which touch both rims. They are a great help, especially when you plan to work with cloisonné. The partitions do not have to be vertical, but they should divide the round form into several sections which can be enameled separately. These dividers are carefully soldered with gold or silver solder after the rims are in place. They prevent the sliding of cloisons and the repair of such misfortunes, which generally take more time than planning the work right from the start.

If a tall shape is part of a tall box, the preparation of the metal shape is similar to the making of a small or flat box. The one place you must watch is the seam. Perhaps you could cover it with a partition, a vertical, which prevents the seam from opening and the solder from causing spots, bubbles, and discoloration.

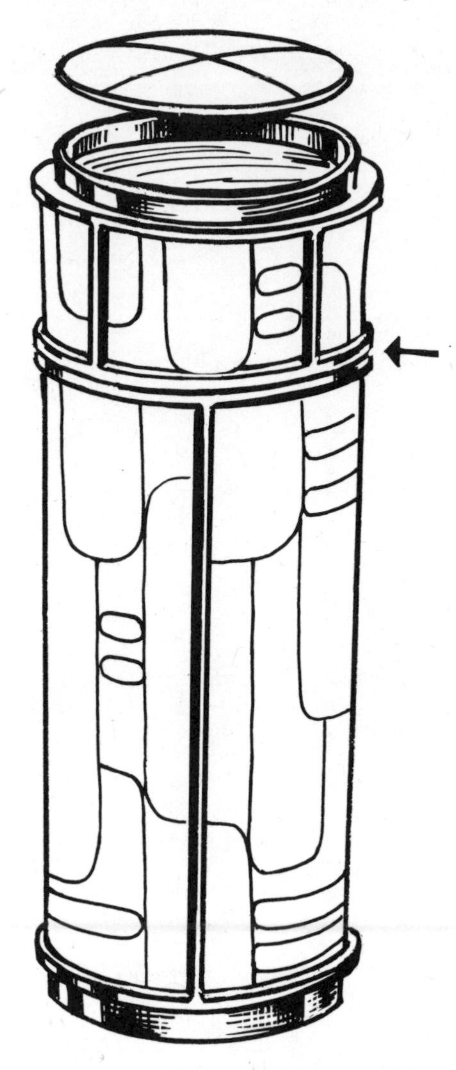

A cylindrical box to be enameled.

THE FOOT OF A LARGE CUP, PREPARED FOR GOLD CLOISONNÉ

The middle part, which is the carrier for the enamel, requires much forethought. In order to have more stability and no soldered seams, I started with a cup shape, taller than the visible enamel would be. This eliminated the distortion of the upper rim of the conical shape which I actually intended. Such a piece would be fired up to twenty-five times and had to stay perfectly round through all these procedures.

1. The first step is to saw a circle of ¼ inch width from a piece of 16 ga. silver, fitting into the open part of the foot at a height of about ½ inch. This is soldered very precisely. This interior reinforcement is important not only during the enameling but also when assembling the chalice.

The next step is to inscribe very exact lines where the upper and lower borders of the enamel are. I soldered gold rims to border the cloisonné. As wasteful as it may seem, I prefer to saw these rims out of a solid piece of gold (18 ga.). The loss is quite minimal, but the danger of solder-joints opening is avoided. When sawing such a rim be sure that the inside diameter is smaller than where it is to fit. One can always increase the inside diameter, but once it is too wide it cannot be reduced. Both rims are cut from the same piece of metal, since one rim is smaller than the other. The remainder of both circles is kept for other work. The rims of course have to be filed slanted to fit the conical shape of the surface; then the pieces are soldered. Always place the solder where the leftovers can be removed most easily — to the outside.

3. Next, the partitions of gold are constructed and soldered with the best possible fit between both gold rims. These verticals prevent the cloison wire and the enamel from sliding sideways in repeated firings. Remember that the enamel has approximately the consistency of honey when it is red-hot. With the supporting partitions we can work in sections and be quite safe.

Chalice with gold cloisonné foot.

4. The last step in preparing the enameled part of the foot of the chalice is to solder a rim around the lower opening of the cup-shape. Use 16 ga. metal, wide enough to provide sturdy support in the kiln yet not too wide, so that a vertical rim on the separate silver footing will meet the inner circle which was attached first.

For further details on construction see pages 62-63.

If I were to construct such a shape of copper rather than silver, I would use 18 carat gold-solder. In working with silver, hard solder ought to be used close to enamels; medium where there is no danger. At all times use a good coat of ochre to protect all joints.

Should the Cup of a Chalice Be Enameled?

A chalice has to take a lot of hard wear and should be well provided with cloisons and metal to brace it against possible shocks and blows. It also must be counterenameled if it is to last.

Most important, enamels contain lead and therefore are poisonous. The use of enamel for dishes and cups is prohibited. I do not think that the small quantity of contamination a priest may be exposed to when drinking from an enameled chalice would hurt him; nevertheless, such is the law. But give a cup a lining of gold or gilded silver and it will be an extremely beautiful and rich chalice.

Detail of gold cloisonné chalice.

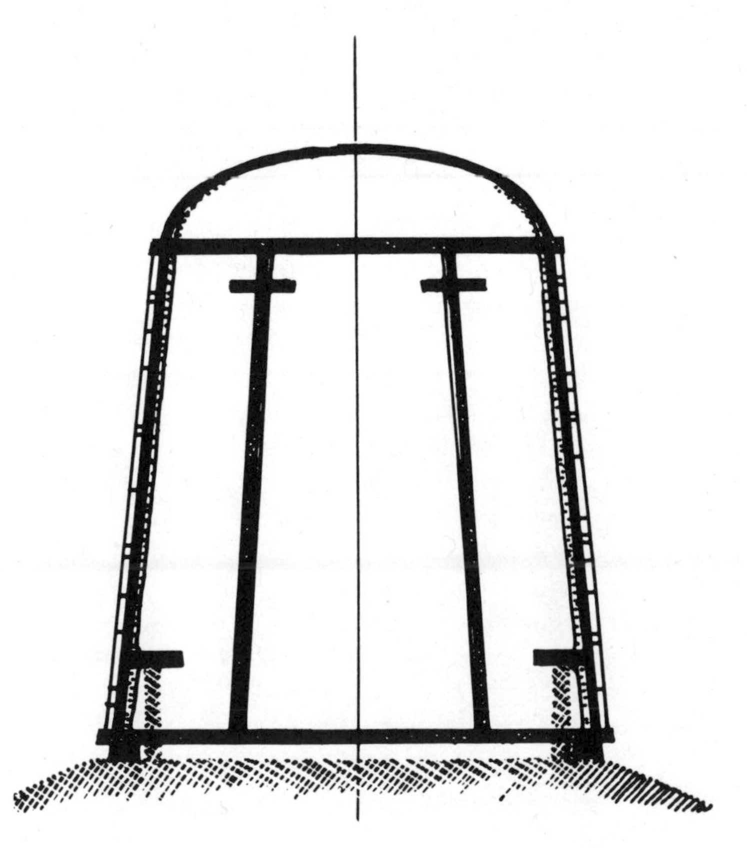

Detail of construction of chalice.

ENAMELED SPHERE AS NODUS OF A CUP

1. Make a pattern of thin aluminum sheet with which to check the symmetry of the two semi-spheres.

2. These two half-spheres are raised from non-tarnishing metal 20 ga. thick, either in a dapping die or over a stake. They must be identical.

3. Have the nut and bolt available which will hold the entire cup together when it is assembled. The size of this *nut* determines much of the metalwork. You may have to change the outer shape of the nut.

4. Solder the two half-spheres on rims wide enough to protect the inside and the outside.

5. Construct of gold or silver — depending on the material you are working with — a small open case for the nut. This case is not soldered to the half-spheres but is to be part of the raised cup, as the drawing shows. But it has to be constructed before the metalwork of the half-spheres can be finished. Cut a hole into the bottom of the case just large enough to permit the screw to pass through.

6. This small case must fit into a collar which will be soldered to a small metal circle (like the islands on the back of pins) right over the center of each half-sphere. The thickness of this circle depends on the thickness of the enamel: 18 ga. for cloisonné, 20 ga. for a thin coat of transparent enamel or with grain-enamel.

7. The height of this collar depends on the design. The sketches explain the making of metal parts and their assembly.

The two semi-spheres are ready for enameling after the usual pickling and cleaning.

8. You will need one more small enamel to hide the bolt at the bottom of the cup. Though it can be seen only when the cup is turned upside down, this piece should be prepared with the same care as any small plaque.

Gold cloisonné sphere with witness marks for exact fit.

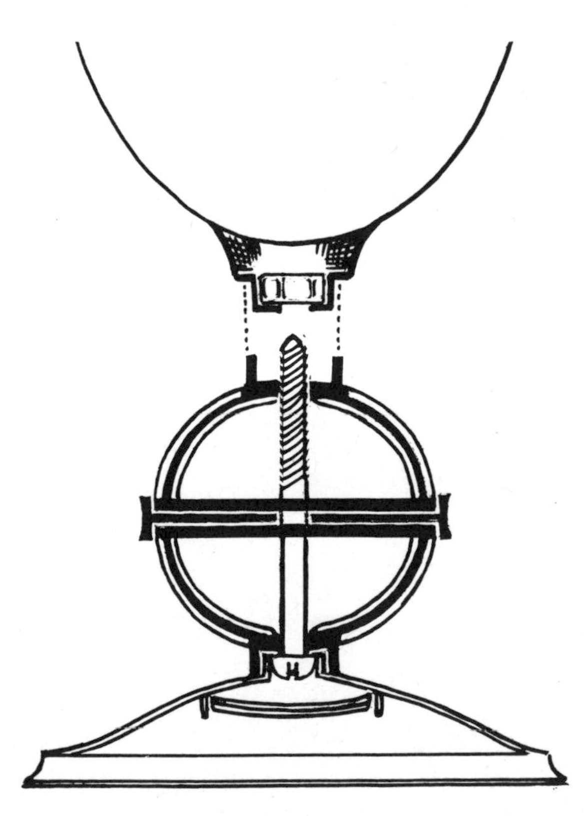

Internal construction of enameled sphere for nodus of cup.

HOW TO PRESERVE THE SHAPES OF LARGE PLAQUES OR FREE FORMS

Let us talk about how to preserve the shapes of large pieces through many firings and how to prepare them for mounting into wood, pewter, etc.

Large shapes to be enameled should not just be big pieces of copper, especially if they have to fit into a wood or metal mounting or are the base of some exquisite enamel.

The danger you wish to avoid is warping. A large round shape might stay well-formed if it is domed and fired on a flat trivet. Still, this circle would be much easier to handle if its edge were turned back, forming a rim of 1/4 inch in width.

For a square or any shape with straight contours, plan the copper base about 3/4 of an inch wider than needed, and hammer this 3/4 inch edge down after cutting off the unnecessary corners, just as you would fold a carton. Solder with medium silver-solder inside a very hot kiln of over 1600°F. When mounting such plaques, drill holes into the 3/4-inch rim, which remains bare of enamel and screw the enamel into its frame.

Frame for a cross.

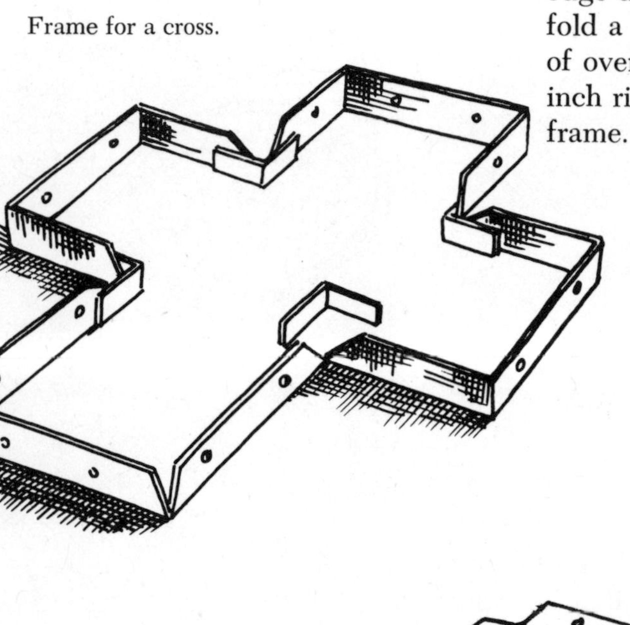

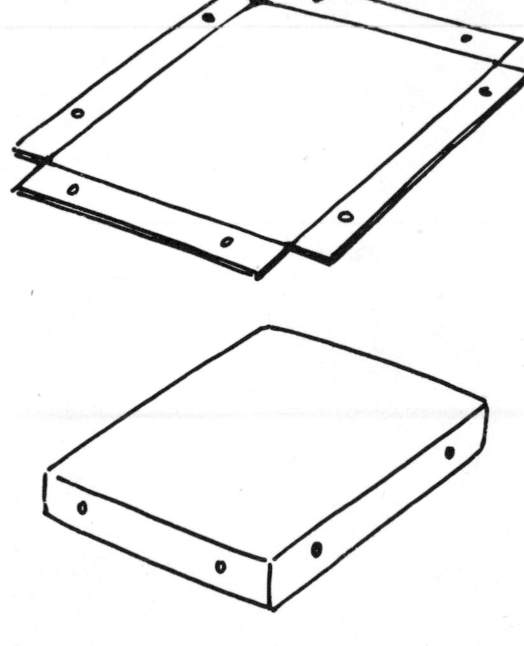

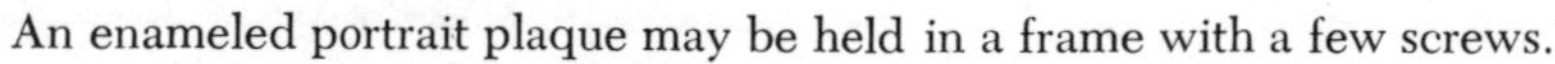

An enameled portrait plaque may be held in a frame with a few screws.

5.

Enameling Techniques

Before you bring the metal shape to be enameled into the enameling area, make sure that the surface is clean, that there is no oxide present, and that all the pieces have been annealed. It should have been in the pickle, brushed with the glass brush, and have no marks or fingerprints on its immaculate surface. Cover the portion of copper over which you will *not* sift with Scalex. The oxide which forms on copper is disastrous in an enamel shop.

ANNEAL EVERY OBJECT BEORE ENAMELING.

SIFTING

If you are sifting a first coat of flux or transparent over a *flat* piece, you may not need to brush or spray Klyrfire first. Sift a thin and even coat while holding the piece at its very edge, turning it slowly under the sieve. Wipe off any surplus that might cover the rim and carefully place the piece on a trivet. If the shape is large, balance it on your left hand and sift from the outside toward the center while turning the piece with the fingertips. Use a bit more around the edges. Then place it gently into the preheated kiln and fire until the surface is shiny and glowing red. It is always better to apply two thin coats rather than one heavy one even if they are the same color. The first coat should be thin, especially if the panel is to be a cloisonné enamel; otherwise, the tiny flat wires (cloisons) might drown in enamel and there would not be enough room for the other layers of enamel. Flux on copper when underfired becomes red and ugly; when fired high enough, it is a brilliant gold with a pink tinge. Making the sample plaque has already taught you which enamels need an undercoating of flux and which do not on various metal bases.

Since we are concerned with enamel in connection with precious metal in this book, sifting and stencils will be necessary only in certain cases for shading larger areas, to get an even background, or for the first coat of a large panel. When sifting washed colors it is possible that these will not stay on the metal but scatter all around. If this happens, you must brush or spray Klyrfire over the base first and then sift over the wet surface. You should also use Klyrfire if the base has a curved shape.

A problem may arise when you need an even transparent coat over copper: the first coat of flux fires through the transparent with thousands of small spots of light. To avoid this, mix the flux and the transparent (1:1) and fire this mixture instead of pure flux. A second coat with pure transparent will produce the correct result.

SIEVES

You need a sieve about 2 inches in diameter and about 80 mesh, and one a little coarser. There will be times when you may want a tiny sieve to reach inside a ring or into the corners of a box. Here is a way to make one:

Cut a piece of tubing (square or round) from a plastic pillbox or a small plastic jar. Heat some very fine brass wirescreen over asbestos and set the plastic cylinder on the wire. Trim off the surplus wiremesh and make a small handle from heavy steel wire (clothes-hanger). Warm the tip of the handle and hold it against the side of the little sieve; the plastic will fuse right away.

If the stencil is held high above the surface, the outline of the design will be diffused.

STENCILS

Stencils are shapes made of paper or metal or an ideal material: old, discarded X-ray films. Stencils for curved surfaces are cut from paper towels and adhere to the metal or enameled surface when moistened with water. When sifting over such a stencil, this shape will have no enamel underneath. If the stencil is lifted carefully, a sharp outlined design will show "negative" in the area covered with enamel. Many advanced craftsmen as well as beginners use this technique which has been described frequently. To achieve a very soft blending of sifted colors in the background of a large panel, I cut a stencil and hold it *high* over the area where I want to keep lighter shades, while sifting dark color. (This is illustrated by the path in the plaque on page 95.) Holding the stencil closely over the object to be enameled produces sharp contours, holding high softens the outline.

What is expressed by cloisonné wires is fact! A shading, however, sifted through (or over) a stencil is a feeling, a hint, a thought, a suggestion – and this could be the most important aspect of the panel.

Stenciling and a sifter.

WETPACKING

Wetpacking means the application of moist enamel with a paintbrush, color beside color. (Don't mix them!) The shading is done by applying and firing several layers on top of one another. Of course the top coats are transparent.

The technique of wetpacking was discussed when I explained the advantages of good color samples. Those samples, of course, were intended to show the properties of one color on different backgrounds. Going one step further and applying one transparent over another transparent, which may already be covering an opaque, results in an unlimited range of colors and shades.

When I explain the three-dimensional effect of the figure of Christ on page 97 or the sleeping men of the panel *Apathy*, you will understand clearly how this is done.

Never *sift and wetpack* enamel at the same time. The water from wetpacking seeps into the enamel and causes water-rims which are permanent.

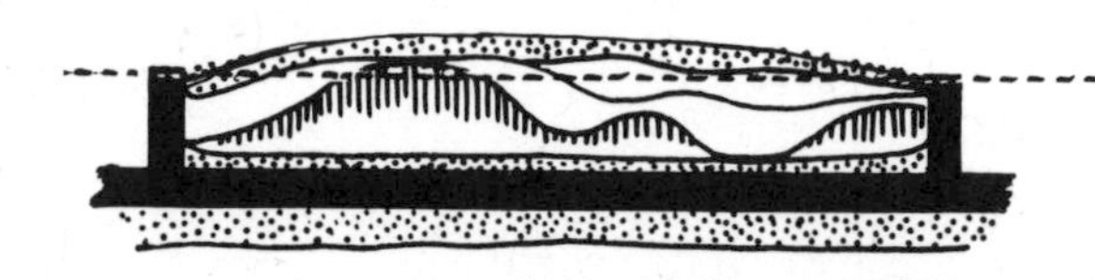

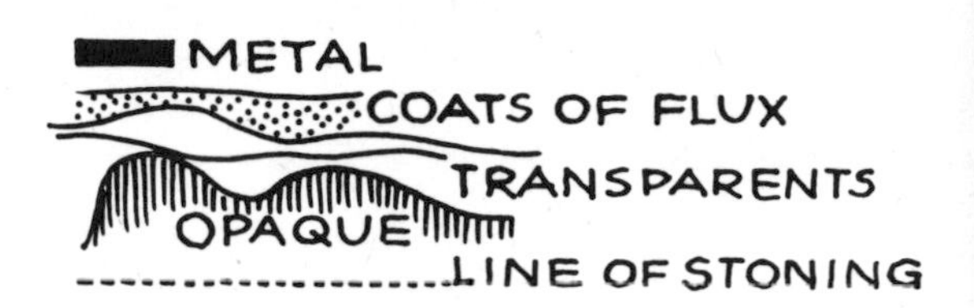

Cross-section of enamels.

Sculptured effect remains when enamel is only fired to a surface appearance we might call "orange peel."

CLOISONNÉ

In the language of the enameler, cloisons are small cells of metal in which ground enamel is fused. These cloisons, or cells, build the design and to some extent protect the enameled piece from blows or shock. Cloisonné work from the Orient and from Byzantium has lasted for more than a thousand years and its colors are the same as they were then.

The design is made of small, flat wire standing on its thin edge and bent into shape. Obviously, a straight line could not stand up, and if we want straight lines in the design we must find ways . . . it is possible! Only overlapping or crossing of wires is impossible. Each wire has to be curved in a way which gives it enough stability to remain in place and to resist the sudden impact of high temperatures. Let's find out which shapes work for cloisonné and which are to be avoided.

The wires form a network over the surface to be enameled. Ideal for the beginner is the repetition of simple forms which have great charm when set side by side. These simple shapes are not only the best exercise for the beginner, but, as the work of Egyptian and Byzantine artists shows, they can be done with superb taste. How lovingly these craftsmen filled a less important space, building protecting walls into their enamels! Not to copy them, but to understand what they did and why, is the true reason for studying these old master craftsmen.

Long curved cells have a tendency to crack, as do places where the wires almost — but not quite — touch each other or the rim of the piece. The ideal shapes for cloisons are compact ones. If the design calls for long and slender cells, break them up into two or three compact ones. The color will unite the design again, and *everything which is right in the technical execution is right in design.*

St. John holding a desperate man: "I came to save and seek the lost." Gold cloisonné enameled in champlevé technique flush on an 18 carat gold panel. (Commissioned by Mr. Chauncy Stillman)

Original drawing from which gold wires were bent for the St. John panel.

Triton and Nymphe. (Owned by Mr. Polk, Westport, Conn.)

Detail showing how much a single line of silver wire can express.

Sizes of Cloisonné Wire

Gold, fine or non-tarnishing:	0.8 mm x 0.10 mm	= 20 ga. x 38 ga.
	or	
	0.8 mm x 0.15 mm	= 20 ga. x 35 ga.
Fine silver:	1.0 mm x 0.20 mm	= 18 ga. x 32 ga.
	or	
	1.0 mm x 0.15 mm	= 18 ga. x 35 ga.
Copper:	1.0 mm x 0.20 mm	= 18 ga. x 32 ga.
	or thicker	or thicker

Byzantine enamels have gold wires of about 0.05 mm thickness, or 44 gauge.

Types of Wire

The various types of wire can be combined if the design calls for it. They can also be used on a different metal — as long as it is either gold, silver, copper, or Tombac.

Gold wire on gold (either fine gold or 18 carat non-tarnishing): This ideal medium needs no flux, does not tarnish, and can take high temperatures.

Gold wire on fine silver: This is beautiful and works well. Blues and greens are brilliant and need no flux, but *reds do need flux for fine silver.* No tarnishing, but watch heat with silver!

Gold wire on sterling silver: Enamel the back first, then pickle. The reflection of transparents increases when the surface is engraved or Florentine-finished. Wires can be set right on the metal. Only *reds need flux for sterling.* The metal does tarnish and the entire surface, except for what remains metal, must be covered with enamel to prevent oxidation. Watch the temperature and use medium or soft enamels.

Gold wire on copper: This works very well. Either use flux for the transparents or use only opaques. It is better to use flux to prevent tarnishing. Transparent reds should be underlaid with thin gold-sheet or foil. Some transparents are very beautiful directly over copper. Since gold wire is rather low, it might be advisable to use such a transparent instead of flux if this fits with the design. If copper is used in small, jewel-like pieces, the rims should be gold.

Gold wire on 18 carat gold which tarnishes: If the gold is an alloy of only fine gold, fine silver, and copper containing no zinc or other metals, this will work. It should be counterenameled first, similar to sterling silver.

Gold wire on 14 carat gold: This will work if the alloy contains no elements other than silver and copper, but it will tarnish almost like copper. Treat it as you would work on copper and make color samples first.

Gold wire on 18 or 14 carat gold containing zinc: Use only if you must and stick to opaques. Small areas may be underlaid with foil.

Fine silver wire on fine silver: This enamels very well. Blues and greens are excellent and only the reds need flux. It is good for panels up to 4 x 5 inches, 18 or 20 ga. thick. Counterenamel first to give stability to this very soft metal.

Fine silver wire on sterling: Use no hard enamels. It reacts in the same manner as gold wire on sterling. Watch the temperature carefully.

Fine silver wire on copper: Excellent for large plaques. Since silver cloisonné wire is higher than gold wire, it leaves more room for several layers of enamel. The base should have a coat of either *flux for copper* or a mixture of this flux and some suitable transparent.

Copper wire on copper: Very good for large plaques. The dark lines of the copper cloisons, especially if treated with liver of sulfur after the plaque is finished, are most expressive.

When you are enameling with copper wire, the piece should be immersed in water after each firing after it has cooled. Brush off all cinder and oxide from the wires until they have been covered by enamel. The cinder might cause bad spots in the enamel if it is not removed. Avoid immersing in pickle. If necessary the piece must be cleaned with ammonia first and then under running water.

All cloison wires must be annealed before you start to bend them into shapes. Otherwise they will be hard to bend and they will change shape when heated. Wind the wire gently around your hand without kinking it. Tie it with two small pieces of steel wire and place it on a clean trivet, mica, or iron-mesh. The tension in the wire will disappear after it is brought to red heat in the kiln. Quickly take it out and drop it into cold water. Gold and 1000 silver wire will be so clean and pliable that they obey the slightest touch of hand or tool.

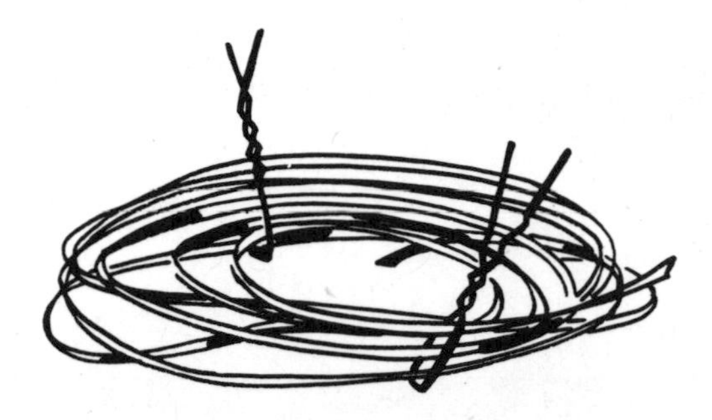

To anneal cloisonné wire, it is held in position with iron wire, heated in the kiln to a red glow, then quickly plunged into cold water.

Copper wire needs pickling and a bath in ammonia-water to neutralize the acid. Then wash it under running water to get rid of all traces of ammonia. I use wires eight to ten inches long for easy handling and cut the wire after the shape has been bent.

If the wires pop off after the first firing of a cloisonné piece, check these things: How about counterenamel? Was the firing high enough to give the wires a chance to adhere? Was the outside coat of enamel too thick and the inside very thin? Was the cloisonné wire annealed?

Small Demonstration Piece

The little piece illustrated is enameled on sterling silver. If it were copper, it would have had a first coat of flux before the cloisons were set. On bases of silver or gold, the wires are glued with Klyrfire right to the metal. This metal plaque was finished with a Florentine graver to increase the reflection and brilliance of the enamel as well as to provide a better hold on the metal. A polished metal surface might reflect beautifully, but it is not as safe a base as a engraved one.

Suggestions for simple designs.

If you plan to enamel a similar small plaque, I recommend that you follow these steps:

1. Draw the outline of your prepared metal base on a piece of paper which you can easily slide under a large piece of tracing paper. You can see the contour and now are free to invent, draw, and sketch. Tracing paper makes erasing unnecessary and you can improve on the design by keeping good ideas from one sketch and using them on the next. This modest bit of material helps to overcome the "respect" for a large, empty sheet of paper; it does not matter how much you waste. It is fun to try the hundreds of ways to fill the small space inside your cloisonné enamel-to-be. "It's only a piece of paper," I like to tell my students. This we later modify to, "It's only a piece of gold." This serene freedom of mind and hand is essential. Try not to draw with shy, little, undecided strokes, but with strong, bold ones which say: That's it!

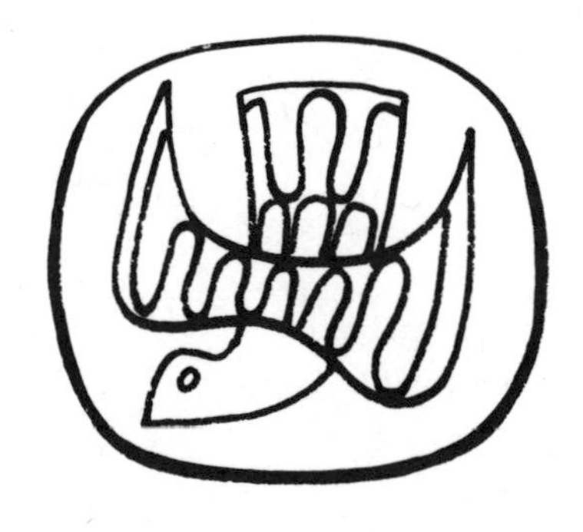

Design for a demonstration piece.

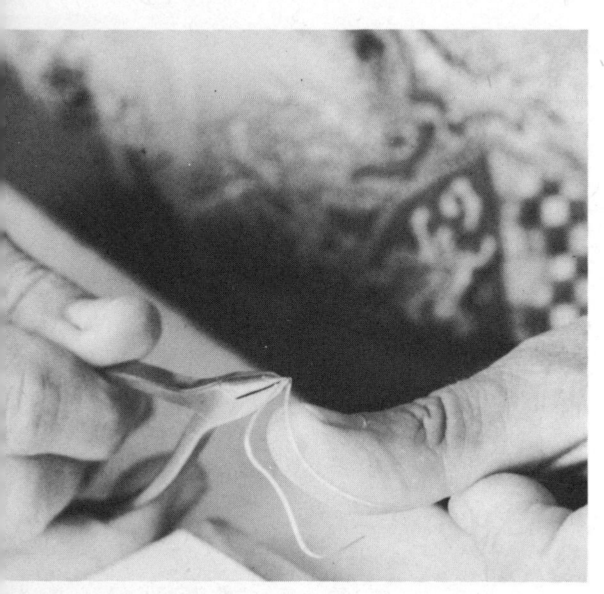

Bending the wires.

2. It is practical to tape the design which you have finally selected under a piece of glass. Stick a piece of double adhesive tape over both glass and design. Every cloison you bend and fit to the design is placed on this tape. Thus you can see your progress and no wire is lost.

3. Shape the cloison wires freely, holding one end of the wire with the special pliers. Don't force and distort the wires with those iron tools! The pliers are only extensions of your fingers. The fingers have all the necessary curves to shape the wire; it is the turning motion of your right wrist which bends the wire into those easy and graceful-looking curves. Sometimes you have to pull a little, while the pliers hold the wire firmly where it is to form a sharp angle, or you may have to push the wire toward the pliers with your left hand in order to get a rounded shape. If you need many circles of identical size, wind the wire over anything round of the proper size – a file, a pencil, etc. – without kinking it, and after removing the wire spiral, cut it open with sharp, fine-pointed scissors. You will have to reshape and close the resulting rings and fit them to the flat (or curved) surface of the base.

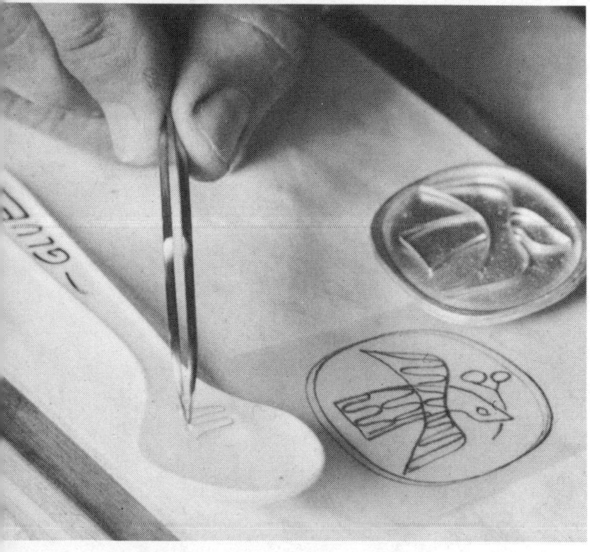

Applying the wires.

4. If you are working on a large panel and have no line design under the flux, find *one* safe point at which to start setting the wires. Find a prominent detail of your work and measure the distance to the top and one side; then mark the corresponding place on your plaque with a "stabilo" pencil. Start setting the cloisons from this point.

5. When you place the wires on the plaque, lift each single piece with fine, polished tweezers, dip it into Klyrfire or tragacanth, and glue it to the base. (The surface on the silver plaque shown needed no flux, while a copper surface would have to have a fired coat of flux on which to set the wires.) Set all the wires in the design.

If the surface of the piece has a strong curvature, the technique of setting wires is a little different. On a background of flux you should add small amounts of enamel to the cloisons at strategic points and give each of these points a drop of Klyrfire. On a bare metal surface imbed the wires very carefully into tightly packed enamel. In either case, fit the wires very thoroughly to the curvature. Before adding the drop of Klyrfire to the enamel, soak out the moisture with the corner of a folded Kleenex tissue. This system will hold the wires very well. If you were to mix Klyrfire and enamel and then apply it, the result would be similar to old stained glass: a frosty appearance full of tiny air-bubbles.

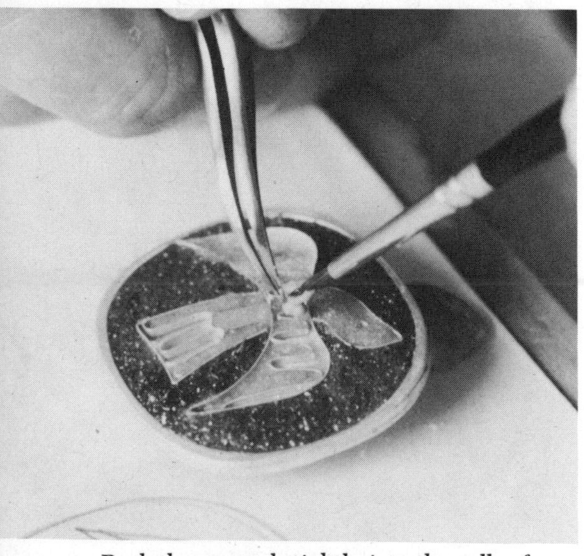

Pack the enamels tightly into the cells of cloison wire.

Klyrfire and tragacanth are glues which evaporate without residues when they are fired. But they hold wires in place very nicely after drying. If you need to make some changes before the piece is fired, drop some clean water from an eyedropper on the area to be changed. In a moment you can lift the wires and move them. Never try to pull out glued-in wires; they will distort. As long as the Klyrfire is moist, you can move the wires and improve your design – a most welcome freedom. Let the piece dry before you apply enamel.

Placing wires on a flat surface without flux is an easy task. Now fill the cells with the first coat of the chosen enamels. The color has been washed and tested on a small sample of the same base-metal.

6. Place the plaque safely on a trivet and let it dry and warm on top of the kiln. Then fire it until the enamel has fused and holds the wires tightly in place. If the plaque had had a coat of flux, you would not yet have added the colors, but inserted the plaque with the wires merely

glued on very carefully into the kiln. When the flux melts, the wires settle into the flux and adhere. A bit of gloss appears at the base of the wires, a sign that they are "in." Do not take such a piece out of the kiln before you are absolutely certain that the wires have settled into the first coat of enamel. The slightest shock would change their positions.

If silver cloisonné wires drown and disappear in enamel, the piece is overfired. There is no remedy.

To gain some experience with this procedure, make some small samples with any bits of cloisonné wire glued on and fire them. The piece will be glowing (not too bright) when you have succeeded.

One of the most delightful moments is when you remove your plaque from the kiln, all the wires in place. It looks so beautiful that you almost want to stop working right here. But there are more joys and more labor to come. You have finished the design portion, now fill in the colors.

The colors, which on the silver plaque are already fired once, hold the wires in place.

May I give you some good advice? Place your work on a small wood block which can be easily turned and handled and allows you to reach every part of your work without touching or lifting it. Cover this block with leather, plastic, or Kleenex so that it can be kept clean and feels comfortable under your hand.

How fortunate that you made the color samples! When you have selected the enamels you will need, grind the transparent ones freshly, wash them, and place them into white plastic spoons. Write the number on the handle and indicate transparents with a "T" and opaques with an "O." If you buy your enamels from several suppliers, it is good to have their respective initials on the handle too. These inexpensive little spoons keep some of the enamel under water while at the tip of the spoon the color is just moist enough to be picked up with either a paintbrush or a small spatula. Use only the tip of the sable brush to transport small quantities of the color to the cloisons; never paint as with oil colors, filling the whole brush.

Before using a different color, dip the brush into the jar of water "for brushes." The grains of enamel will settle to the bottom and you can continue wetpacking.

Transparents are exciting and should be treated with restraint, like delicacies. If cells are small, transparents can have the effect of precious stones in a small room after dark: they simply don't show, while opaques will present intense color right at the surface. On the color samples opaques might remind you a little of ceramic kitchen-tile: too shiny and too flat. But if you fill them not quite to the top of the cells and, at the next firing, cover them with a coat of clear transparent of the same color, or cover light opaques with clear flux, your opaque suddenly has the warmth and charm of alabaster. Later on, when the piece is stoned smooth and waxed with the best paste-wax, you will agree that opaques have a dignity which transparents often lack. The Byzantine master knew what he was doing; he used transparents very seldom – only in a halo.

After your plaque has cooled, add the second coat and fire. Perhaps a third coat or more is necessary while you are shading and changing the hues with each filling and firing. Proceed until the cells are completely filled with enamel.

The "rippled" tool moved along the side of the plaque evens the wet enamel and brings the water to the surface. . .

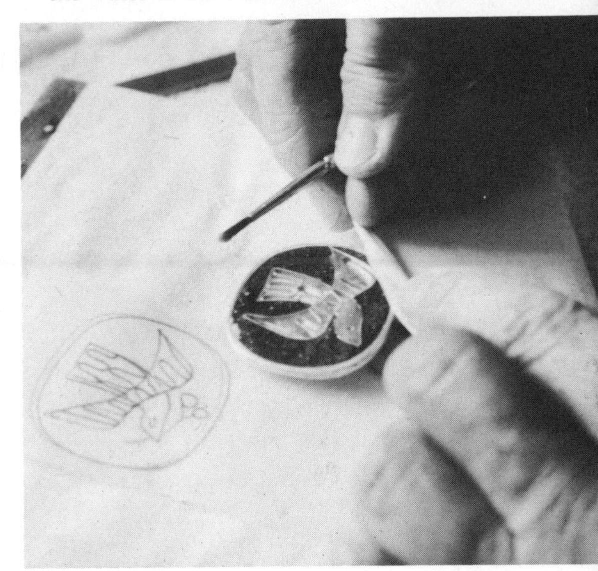

. . . where it can be easily removed with Kleenex.

The plaque is dried and fired.

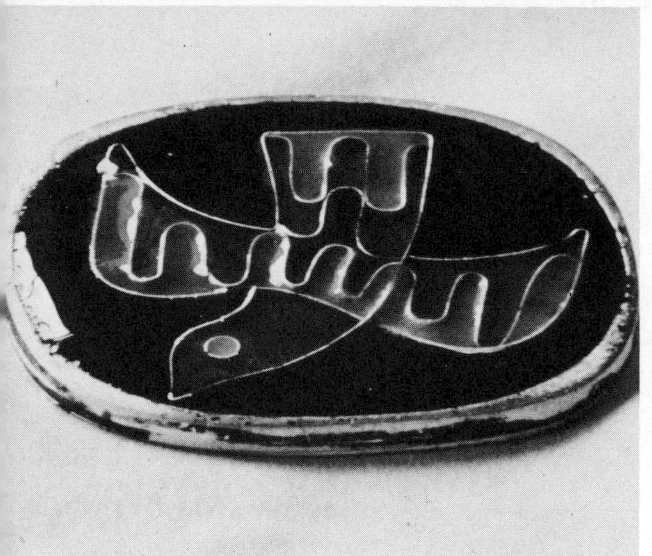

The plaque is fired and ready for stoning.

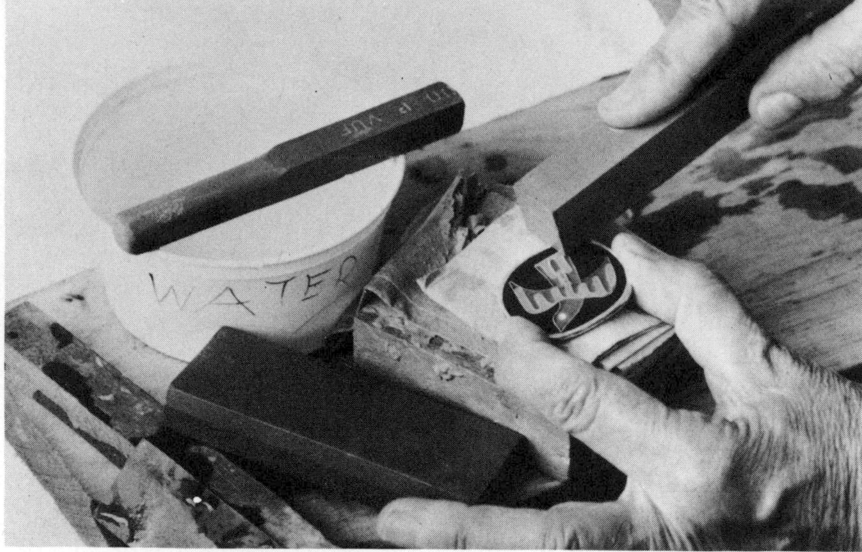

With wet carborundum stones, medium and fine, and scotchstone, the enamel is stoned until all cloison-wires are free from enamel and the surface is perfectly even.

Stoning Cloisonné Enamel

Stoning should be done next to running water. Use a medium carborundum stone at the beginning and rub it over the surface in one direction, rinsing frequently. Change the direction of stoning at times in order to reach every place and to get an even surface with all wires exposed. When stoning rims of bowls or plaques, slant the stone from the inside outward, or the enamel might chip. There will still be some depressions, still glossy, while the stoned areas look rather discouraging.

Now take a finer carborundum stone and work the entire surface over once more. Rinse the gray silt away often; it causes scratches in the surface even while you are stoning. Then go over the surface with medium wet-and-dry emery paper, using it wet. Rinse again and check that all wires are free of enamel. When no more stoning is necessary to show the wires as shining metal lines, rinse, wash, and then use the glass brush under a jet of water to remove every little impurity left by stoning. The gray silt will come right to the surface when the piece is fired again and cause ugly gray spots that can only be removed by stoning or with diamond-impregnated wheels or points in a flexible shaft. Clean with the glass brush before refilling and refiring. When the plaque is thoroughly clean, take it back to your wood block and fill in all low and glossy areas with transparent enamel in the case of a dark shade, or clear flux if a light area has to be filled. Flux over a dark color fires milky. Should there be any pores or holes, open them first with a pin or a diamond-impregnated wheel in a flexible shaft. Chisel out bad spots with a sharp tool. Then wash and refill the surface with the enamel, making small mounds since they are reduced greatly in firing. Pass over rough stoned places with a wet brush and a little color, then dry and fire the piece again. The surface looks much nicer now, still a bit too shiny and not yet perfectly smooth. This time stone with soft carborundum and then, when the entire surface is matte and smooth, use the wet-and-dry emery paper (wet) until the surface is as smooth as a baby's cheek. These last twenty minutes of patience make all the difference between good work and better work.

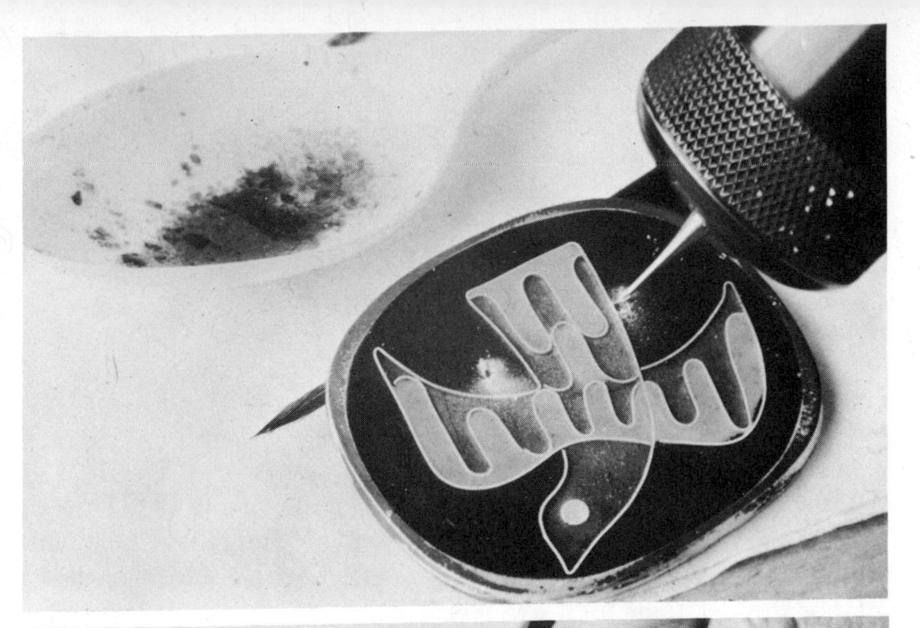

If there should be some pores or impurities, they are opened up and ground out. After thorough washing under running water, they are refilled, refired, and stoned. The finished piece is then either flash-fired or waxed with a good paste wax and hand polished with cerium oxide.

A highly polished surface is achieved without refiring by rubbing the well-stoned surface with felt and cerium oxide (both wet). This polish is as smooth as a precious stone.

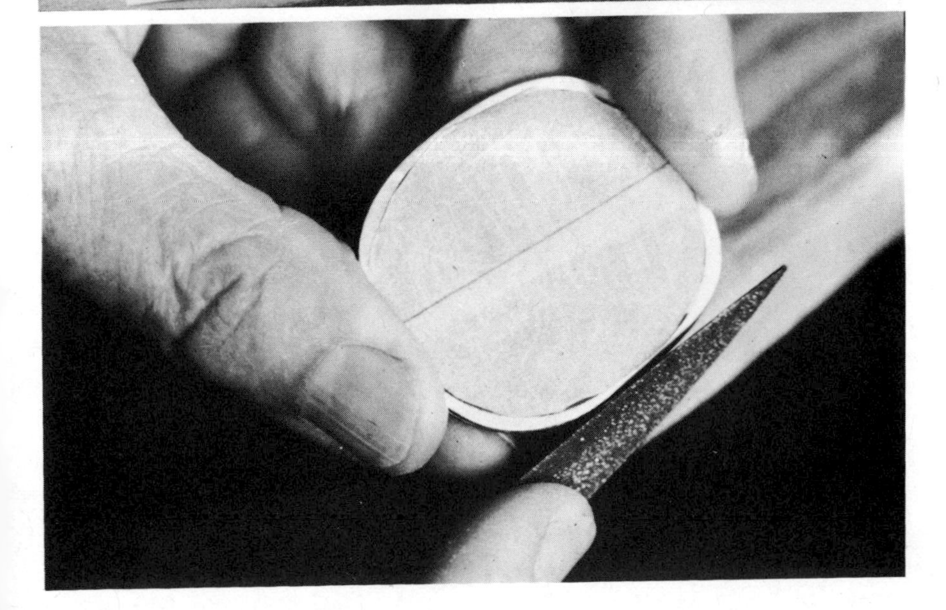

Filing and finishing the metal. Cover the enamel with tape for protection.

Finishing — Three Ways

You now have a choice between three different finishes.

The matte surface

All this involves is rubbing in one or two coats of the best floor or furniture paste-wax available. Be careful — even the lines in your fingertips leave marks. Let it dry halfway and then rub in the other direction; then leave it alone for a while before you polish. The best tool for polishing is the ball of your hand; its temperature and texture are perfect.

A shiny, glass-like surface

Flash-fire the piece fast and at high temperature, possibly upside down. The effect will be smoother.

A highly polished surface

Place the piece on a block of wood cushioned by paper towels. If you drill a few holes into this block and have some wire-brads (nails without heads) sticking out of it to keep the enamel from sliding, you can make the polishing much easier. Dip a piece of thick wet felt (a buff) into moist cerium oxide and rub the surface of the enamel for about ten minutes in all directions. A large or three-dimensional piece such as the bishop's ring takes longer to polish, but the result is worth the effort. Your plaque will be as smooth as a precious stone and should be set and treated like one.

If you did a good job with this first piece, you are no longer a beginner. You are now ready to enamel a band ring, rich with cloisonné inside and outside.

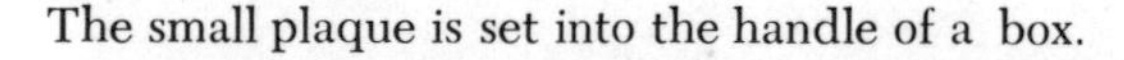
The small plaque is set into the handle of a box.

Beginners enameled these three rings.

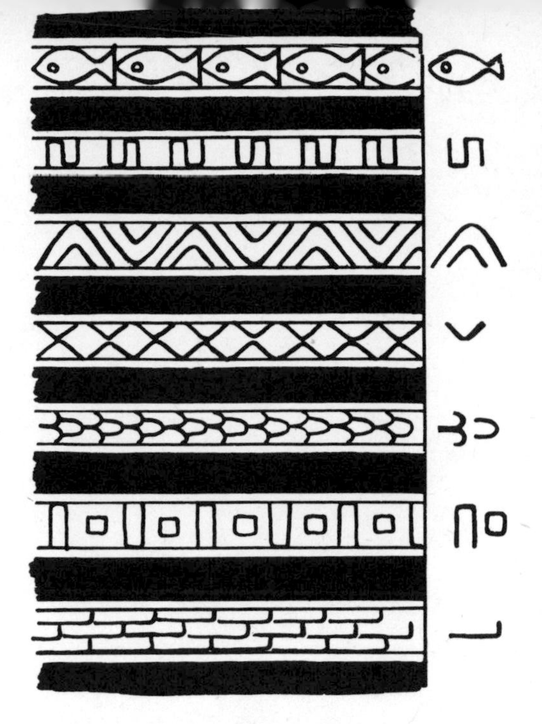

Designs for band rings should be as simple as these.

CLOISONNÉ ON VERTICAL SURFACES

The Band Ring

The construction of the ring is shown on page 38. The design should be kept simple. It could be lettering or small arches or squares alternating with vertical rectangles. Look at Egyptian jewelry and you will know what I try to express. Long waves all around the ring are nearly impossible to apply; they would have to be made in many small segments. The design should touch the rim in at least four spots, to take the place of soldered partitions.

Let me tell you of two different ways of enameling band rings: the first is the approach for an adventurous character, the second is more suitable for a lover of intricate designs. In one you wetpack the entire inside and outside, cloisons included, at the same time, fire it, wetpack a second coat, and if all goes well, stone and finish the piece.

In the first method, spray or paint Klyrfire over the inside of the ring and sift a thin coat of enamel, first from one side, reaching especially the niche where the rim is soldered to the band. Then turn the ring and sift from the other side. If necessary, "salt" with your fingertips where enamel is missing. Spray some Klyrfire over the entire inside, then daub with your little finger over the inside, pressing the enamel even and tight and wiping the rim. Dip any cloisons you might want inside the ring into Klyrfire and press them gently into the enamel. A small drop of Klyrfire will help hold the cloison. Let it dry and clean all grains of enamel from the outside. Then set about one quarter of the outside cloisons, holding them in place with Klyrfire *and* enamel. Soak out the moisture and add a drop of Klyrfire. When the enamel covers this portion of the ring with a good, tight coat, proceed to the next quarter of the ring. Be careful not to use too much water and soak out all moisture before adding Klyrfire. Too much water will loosen the previous segment. Continue to the next in the same way until the entire outside is done. Let it dry and place it flat on a piece of mica or a trivet. Then fire the ring fast and high. If the enamel was too thin, there will be bare spots. If it was too thick, it might fall off. If it was just right, you are ready to stone.

Sifting enamel on the inside of the band ring.

In the second method, sift over Klyrfire a thin, even coat of transparent enamel on both sides and fire the ring. Place the ring so that you can see part of the inside and part of the outside simultaneously. Set the wires in four sections, inside and outside at the same time, securing the cloisons with some enamel at strategic points. Add a drop of Klyrfire and fire each segment of the ring separately, just to hold the wires in place. Now you can take all the time and care necessary to achieve three-dimensional effects, to underlay with gold, to create textures and shadings. The subsequent firings can be done separately for each segment of the ring, or the final firing, which will only serve to fill some low spots, can be done with the entire piece lying flat. Such a ring should be stoned and polished very carefully.

The cloisonné fish are applied with Klyrfire directly onto the silver surface.

A band of stainless steel can be bent to serve many purposes.

The cloison wires are embedded in enamel on both sides of the ring.

Enameling a Box

Having had experience with band rings, you already know how to enamel boxes – inside, outside, on top, and underneath. Don't forget the counterenamel of the plaque which is to be set into the cover. One suggestion: the enamel on boxes or any other shape with rather sharply bent corners tends to crack right at those corners. If some soldered verticals are included in the design, the sections will be relatively easy flat areas. The seam inside presents a small problem. It will be visible under transparent enamels. There is nothing objectionable about a neat seam, but if you wish to conceal it, plan to use some cloisonné or other design over it.

A finished box.

1. A simple way to hold the box while applying the wires and color. 2. The trivet is bent so that it touches only the metal rims. 3. Gold cloisonné covers the inside seams.

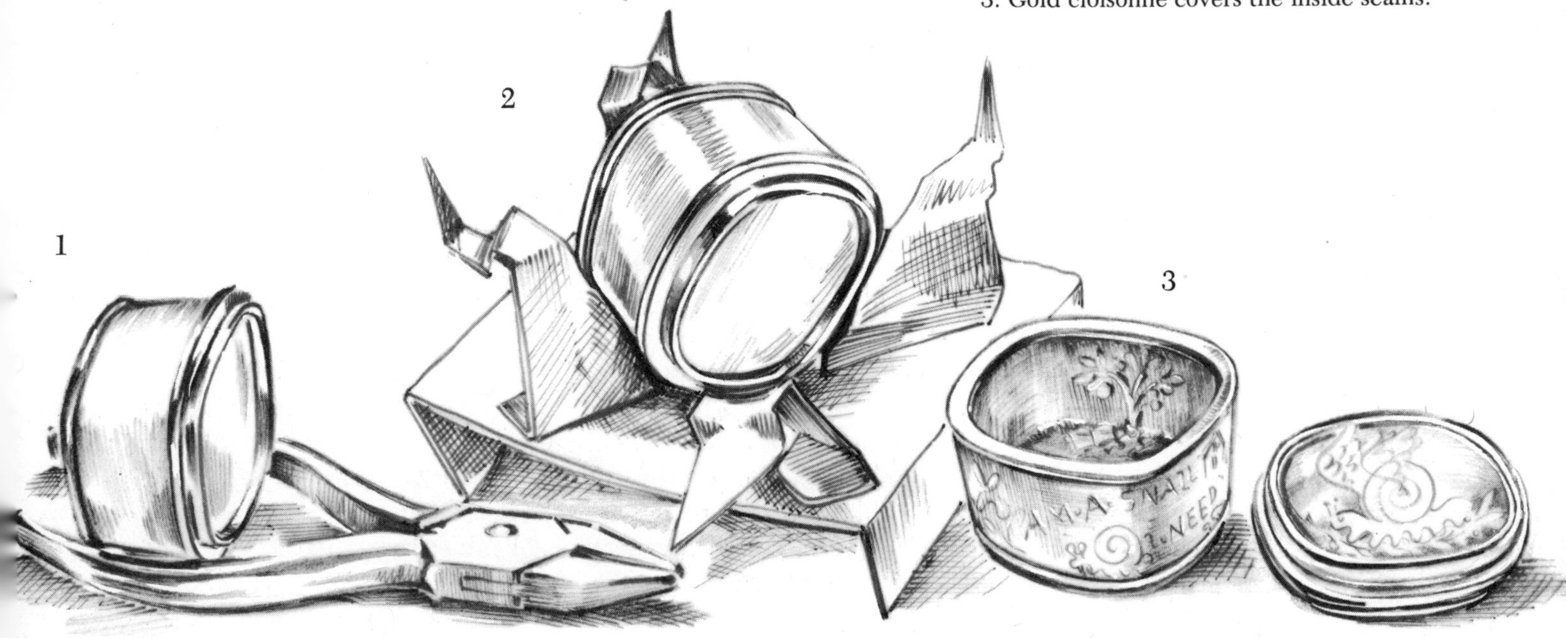

Enameling the Foot of a Chalice

The construction of the metal parts was discussed in the chapter on "Metal Shapes for Enameling." The opposite page shows sketches done in preparation for the making of the chalice and while the work went on. They show clearly those small skills and details which are so easily forgotten but are so essential for successful work.

1. The exact conical development of the enameled part of the foot. With its vertical divisions it is the skeleton over which the composition of fourteen figures had to be designed.

2. The first inside and outside coats of enamel had been fired. The piece was ready for the setting of gold cloisons in each section. To have easy and comfortable access to each point of the work-area and to prevent any rolling or moving of the piece, I made three small cushions from Kleenex and glued them with tape to a small wooden board, also covered with Kleenex, on which the cup-shaped foot rested safely.

3. A large diamond was to be set into the silver base of the foot, right under the crucifix. This setting is made from 18 carat gold with a platinum rim to hold the diamond. The small bowl of gold holding the mounting is highly polished for good reflection. The stone had to be set deep for reasons of taste and protection but so that it would get enough light. After completion of the small gold unit it was soldered into the silver base.

4. All parts being finished and polished, the assembling of the chalice still held some challenges. The silver base, many times heated for soldering, had to be checked for perfect flatness before the enameled part could be added. A flat piece of hardwood with a hole fitting exactly over the collar of the silver base was hit with firm but gentle strokes, using a wooden hammer. The wood block was held on the opposite side and the hammer tipped slightly to the inside, hitting vertically only.

5. Leveling and centering the three parts on a turntable: Masking tape holds them in position, while the scriber marks the fit and especially the points where the enameled part (crucifix) would be.

6. The nut is encased and bolted in so that it can neither move nor fall out during the soldering procedure (upside down). The cross witness-marks in front indicate where the crucifix will be, leaving no gaps when both parts are soldered together. The heating is done very evenly to bring both parts to the same temperature (color of glow), and to secure a good flow of the silver-solder.

6a. The shape which is applied to the center of the cup.

7. The screw head in its recessed small silver box, inside the foot. It will be hidden under a small gold cloisonné, which is set like a precious stone and can be removed if ever a separation of the parts should be necessary. Just unscrewing this one screw makes repairs or re-gold-plating of any part possible without damage to the other parts.

7a. Three small blocks of silver are soldered to the inside of the enameled part above the inside rim. They help to resist pressure, which might be applied if the cup ever has to be re-assembled.

The work-drawing shows the conical development with five verticals, three crosses, and two outlines of the crucifix, placed in such a manner as to divide the whole band into more or less equal parts. The five golden verticals were hard-soldered and they had to be included in the

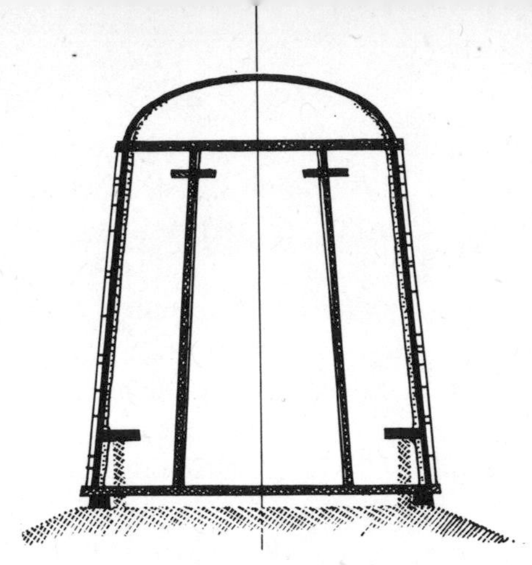

Steps in enameling and assembling the chalice.

Drawing for bending gold cloison wire for the chalice.

Detail of foot of chalice. The cross is actually one of the partitions, soldered to the base before enameling.

design with fourteen figures. The first rough sketch puts on paper approximately what is intended. Then, drawn on tracing paper (one on top of the other), the individual sections were developed, but not yet detailed. It is much better to "draw" with the wires while bending them. The design grows more naturally and comes to life, much more so than a slavish repetition of the pencil work.

First a coat of flux and white (mixed 1:1) was sifted over the exterior. Prior to firing, those parts which would not be enameled (the rims, the verticals between the edges, and the metal above the rim) were cleaned of all enamel, since some always settles in places where it should not when you sift. Then came the *first firing.*

Next, the interior was carefully counterenameled. Special attention had to be given to those places where the rims had been soldered. (If the sieve cannot reach these areas, sprinkle enamel with your fingers and tap it firmly into place.) Then came the *second firing.*

At this point I was ready to set the cloison wires, holding them with bits of enamel where they might slide, especially on the sides of each section. When one section was finished, I fired the wires onto the whitish first coat. White was chosen to give intensity to the colors when using opals or dark transparents. Brilliant transparents over metal in this very serious design would be too festive. Opaques were used mostly, and they were covered with transparents to give them depth. Since I used gold wire only 0.8 mm high, this first coat of flux and white had to be very thin but cover the surface well. There had to be space for three or four coats of enamel in each cell to give shading to the figures. Since there were five sections, *five firings* were necessary before the wires alone were firmly set. Beginning on page 95 is a description of how shadings and three-dimensional effects are achieved.

Even though the figures on the foot of the chalice are small, they were executed in the same manner as large figures.

After all cells were filled and stoned, I decided on a matte finish, stoned to the greatest perfection and smoothness. Rubbing with scotch-stone and wet-and-dry emery paper, used wet gave the piece a silky appearance. Later, I used a fine paste-wax and repeatedly polished with the ball of my hand and clean woolen rags. This surface promises that the enamel will be more and more beautiful through the years, since the hand touching the enamel will have this same bit of warmth and wax as the hand that made it. A flash-fired or highly polished surface could not become better with age, but would turn slightly matte.

Parts of chalice before assembling.

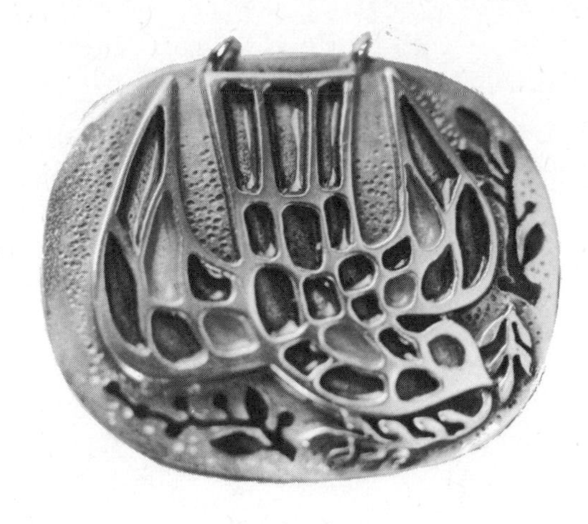

Concave enamel. Height: 1¾ inches.

CONCAVE ENAMEL

Concave enamel is a variation of cloisonné. The wires are more substantial, about 0.5 mm thick, and the enamel is very clean and transparent. It is wetpacked over bare metal (silver or gold) and built up high at the rim and the cloison-wire, and leaves just a thin, even coat in the middle of each cell. All cloisonné wires must touch, forming small cells. When fired high, the enamel raises up along the wires and leaves the middle low and reflecting like jewels. When stoning, do not touch the enamel with the stone; only the wires are to be cleaned of enamel, then filed and polished with scotchstone and a chamois polishing stick with rouge.

Should the enamel accidentally be scratched by the stone, clean it with ammonia and water and refire the piece. The bird shown here as a sample of concave enameled cells is champlevé, not cloisonné.

Wetpacking cells for concave enamel.

St. Christopher. Champlevé. The enameling is done inside the low fields *(champs)*, then stoned and polished to perfect evenness with the surrounding metal.

CHAMPLEVÉ

The word "champlevé" (raised plane), one plane above the other, indicates how the metal is prepared for this ancient technique of enameling. The enamel is then applied and fired into the openings of the upper plane. When it is finished, the enamel will be flush with the surrounding metal.

There are several ways to prepare the metal for champlevé enameling.

1. The first is to engrave the depressions, to chisel them out of the metal. It takes a skilled craftsman to do this. When such pieces are enameled and fired before any oxide has caused tarnishing in the engraved areas, the colors will reflect exquisitely. (Antique Indian jewelry is a wonderful example of this.) The depth for the grooves is between $\frac{1}{30}$ and $\frac{1}{50}$ of an inch, depending on the number of coats of enamel. The technique of enameling is the same as cloisonné; the two techniques can very well be combined on one piece.

Since traditional techniques are not inflexible laws which all must obey, let us develop this further and go on to our own experiences.

2. We can etch the grooves rather than engrave them. The whole piece except for the top has to be covered with asphaltum or wax to prevent the acid from etching in places where it must not etch. If the design is intricate, paint a coat of asphaltum over the front, after the back and the sides have dried completely. The asphaltum takes about one hour to dry to the point where it can still be inscribed with a dull needle. Larger areas can be removed with a hollow scraper or a fine knife. Before etching, clean off the particles of asphaltum which are a by-product of the incisions of the design. Rub steel wool gently over the surface to get rid of accumulated asphaltum. The inscribed lines must leave bare metal clearly visible, with no asphaltum preventing acid from reaching the metal. The asphaltum coat must have no thin areas or open spots. Acid will eat in such places.

If the design consists of bold, clearly separated areas to be etched, the asphaltum may be used like paint, covering the portions to remain as they are, leaving exposed the areas to be etched.

We can use either nitric acid on copper and silver, or ferric chloride for copper only. Nitric acid, mixed 2 parts water and one part acid (water first!) will etch in less time than ferric chloride, but the lines will not be as clear as with ferric chloride. Nitric acid etches in all directions deeply and widely. Check after one half to one hour.

Ferric chloride, mixed 1 part acid to 1 part water, takes a longer time to etch. It eats vertically into the copper, keeping the contours cleaner.

All etching must be done in a well-ventilated room, possibly a separate small place. The fumes attack all metal surfaces and cause tools to rust, not to mention their danger to humans. Cover the acid bath; uncover it only when you need to remove the small white bubbles over the etched areas with a feather. These bubbles retard the process. To measure the depth of the etched areas, use a sharp wooden stick while the plaque is still in the acid, and use care not to damage the asphaltum.

When the necessary depth is reached, remove the piece from the acid, rinse it well, and clean off the asphaltum with gasoline. Then immerse the piece in pickle, heat it, rinse again, wash with ammonia, water and pumice, rinse again in water and dry the piece. It is ready for enameling.

Here is another idea. Why not paint only a rim of asphaltum around the back of the piece, provided the metal is thick enough, and *etch* the area for the counterenamel while the front is etching? You may turn the piece in the acid bath to have access to the other side.

Remember, when you are working with copper, that all areas not to be enameled must be covered with Scalex before each application of the next coat of enamel.

Cleaning bubbles from an etched area.

When the enameling is done, the stoning must be done with extreme care. You don't want to scratch the framing metal and later be faced with the hard task of polishing the metal again — by hand.

There might be some pores or small flaws showing while you stone. Go on and finish the metal to perfection, even polish it. Yes, the piece must be repaired and refired, but the polished metal offers less chance for oxide to develop. After the last firing is completed, you need only go over the metal with a chamois stick and rouge to repolish it. You may have to dip the piece briefly into pickle, not long enough to damage the enamel.

3. If you are not an engraver, there is a good technique for very fine champlevé with precious metal, gold or fine silver or a combination of these, gold being the top metal. Saw the design out of gold (18-20 ga.); very fine shapes can be cut out with good sawblades. Solder the two sheets of metal together, placing the solder where it is least likely to interfere with enamels and where it can be reached with a file. Solder a rim around the back to provide for counterenamel. Add gold cloisonné to the front if you like to combine the two techniques. The results are very beautiful.

4. A fourth way to prepare metal for champlevé might be of interest to the sculptor-craftsman. Make a model of any shape from wax. The inner metal in the lost-wax cast must be thicker than 1/32 of an inch and it must be counterenameled. If a round sculpture is to be enameled, it should have an opening at the bottom through which a thick mixture of Klyrfire and counterenamel can be poured. (Swish it around to cover the entire inner surface and pour off the surplus.) This opening may be sealed with a piece of metal after the work is finished. The piece will have to be stoned and polished and therefore must be accessible to tools. This wax model is then to be cast in the "lost-wax method." Enameling will then be no problem if you understood the discussion about enameling tall vertical shapes. Metals for *casting and enameling* are either Tombac or sterling silver.

Such a sculpture may be gold-plated or darkened with liver of sulfur and then rubbed with paste-wax and a woolen cloth.

St. Peter. Cast in Tombac for champlevé.

PLIQUE-A-JOUR

Plique-à-jour is enamel in a structure of precious metal, with the charm of stained glass windows set in lead and stone. Light must shine through it, or it loses its meaning. This fact limits the number of applications for *plique-à-jour*. In earrings and hair ornaments it is appropriate, as well as in religious objects such as a processional cross, an eternal lamp, or the crowns of a thora.

I made a round small sample piece in order to describe the technique step by step and the piece was mounted into the handle of the concave, highly polished top of a pewter box. The reflection from this concave form brings the *plique-à-jour* to life, changing with every difference in light or perspective. It does more. Looking at the box, one is fascinated by the enlarged reflection in the cover; something happens all the time — a play of color and light where actually there is "nothing but a piece of polished pewter."

The design is made on tracing paper glued on the silver; holes are drilled where openings are to be, then sawed out so that the metal skeleton for the piece has *no* soldered joints.

The piece is held tightly over thin mica on fire brick. Hammer small tacks into place where necessary to keep it down.

Fill in the freshly ground enamel, heaping it as high as possible over each opening.

When fired, the enamel will have melted quite deep into the openings. Stone the metal parts free, then refill and refire until no hollow spots are left.

Stoning both sides over a piece of thick felt to soften the impact of the stoning.

If, after the little "ears" are sawed off, some repair work is needed, tie the piece with iron wire tightly over mica and fire brick, not touching the enamel, only *metal.*

Refill generously and refire.

The enamel is finished and set into the handle of a box, producing colorful reflections on the polished, concave metal.

The design should provide openings of not more than ¼ of an inch by ⅜ of an inch. These ought to be compact shapes similar to those used in concave enamel.

This piece is made from one sheet of 18 ga. fine silver. I prefer sawing out all the cells to using any solder with *plique-à-jour*. Wherever there is solder, there is the danger of discoloration or a soldered joint opening just enough to cause cracks and give trouble. Also with one solid piece we can have a good rim and can leave small "ears" around the outside which help to hold the piece firmly to the mica and the trivet. It is possible to saw so neatly that nearly no filing is necessary. If the sawblade is held a little bit slanted, the cells will be conical and the enamel has a better hold. The roughness caused by the blade also helps.

The design is again made on tracing paper which is glued over the silver. Check the size of all intended openings once more, then drill a hole into each cell to permit the sawblade to pass through, then cut out the cells. Anneal the piece and at the same time burn away the paper and glue.

The silver remains white and needs only to be washed and dried. A copper or sterling base has to be boiled in pickle, washed in water and ammonia, then again in clean water, and brushed thoroughly with the glass brush.

The next very important task is to hold this piece firmly on a flat slab of fire-clay, with a thin layer of mica between enamel and fire-clay. Mica must be thin because it blisters and may cause unevenness; therefore, slice a layer off from used mica rather than new mica. Shape and cut small staples (U-shapes) from medium-strength iron wire. These give a very sturdy hold when hammered into the fire-clay over each of the small "ears."

Enamels for *plique-à-jour* must be freshly ground, not too fine (60 mesh or even coarser) and very well washed. They should look like clean sand of precious stones. Pack them as tight as possible. It is amazing how much enamel can be put into one little cell. Pack them with a small mound on top, being careful not to cover the partitions. Give each packed cell a drop of Klyrfire after the moisture has been soaked up with tissue. Dry the piece and fire it fast and high. The enamel crawls to the middle of the cell if underfired. If it is fired very high, it melts up the rims, forming a nice flat coat of color. Some cells may have holes; all need to be refilled. Wetpack tightly and leave mounds on top. This time you may not need glue, but dry the piece carefully before firing a second time. In most cases two firings are sufficient.

Throughout this time, the base has been held firmly to the firebrick. Do not remove it until you are certain that all the cells are well filled. There will be a little enamel on the metal portions of the back, some mica still sticks to it. There might be places in the front where the enamel has surface cracks, usually where it covers the metal bridges. Before you start stoning, fold a paper towel like a handkerchief, lay it on a wooden block about the size of your work, and put the *plique-à-jour* on this shock-absorbing, wet pillow. Don't stone the back yet!

There might be some defects, pores, cracks, depressions, or even blemishes which should be taken out again. In such a situation you must make a hard decision. Rest the piece on leather or paper and chisel the bad spot out. It is difficult to say what kind of tool to use; you may have to make one out of an old file.

When this is done, wash and clean the piece thoroughly under running water, turn it upside down, and place it again on a thin sheet of mica held firmly by the staples. Again fill the cell tightly and generously from the back. The mica will stick to the front, but you have to stone anyway, and the enamel will be held in place by the mica. Fire fast and high, very high to preserve the clarity of the enamel and "to catch the piece by surprise" before the enamel can drop through the holes.

There will be little to stone again on the front; the back will also be quite even. Both sides may need some gentle carborundum stoning, wet-and-dry emery paper, scotchstone and one of the surface finishes described on page 58.

Let's pretend that there is still a flaw and the piece must be refired but the metal is already polished and the "ears" have been sawed off. How can the piece be held tight enough to mica and fire-clay? Tie it on with iron-wire as tightly as possible, but without letting the wire touch the enamel. Refill, fire and proceed with the finishing. Or refill the damaged place and add a small amount of enamel to each cell, forming flat mounds. Put a drop of Klyrfire on each and keep the metal free of enamel. Then find or make a trivet which permits the piece to be fired at a 45° angle. Back the piece with mica where it leans against the trivet. Again fire fast and high — very fast this time, and watch for the moment when the enamel has fused. Each cell will be slightly convex — which is very good. If you hadn't added enamel to all cells, there probably would be several depressions in the surface now.

If the enamel has already begun to drop down, take the piece out of the kiln, turn it upside down very quickly and refire. The enamel will slip back into place. This piece should remain flash-fired.

A nice addition to *plique-à-jour* are small gold balls or gold cloisonné fired into the "windows" of translucent enamel.

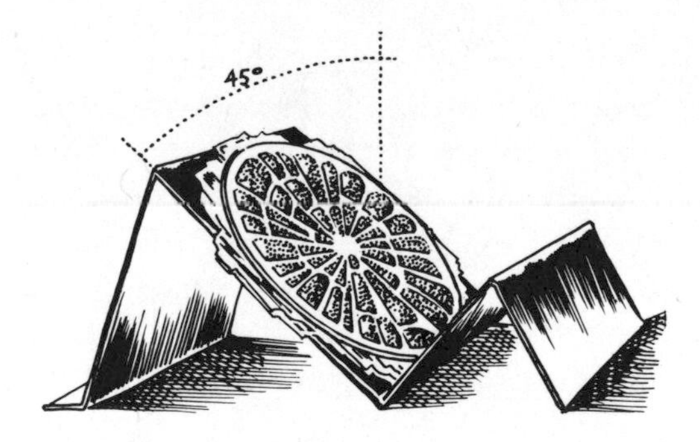

A way to make minor repairs. Bend stainless steel so that enamel rests at a 45° angle over a thin sheet of mica. Repair and fire very fast and high. If the enamel drops slightly to one side, turn the piece and fire again; it will run back.

Plique-a-Jour On a Curved Surface

Another method for doing *plique-à-jour* enamel on both flat and curved surfaces is as follows: Roll copper very thin, no more than 1/10 of one millimeter, and pickle it after annealing. Glass brush the front and paint the back with Scalex. This piece of copper should be the same shape as and about one quarter of an inch larger than the planned *plique-à-jour*. Sift a medium-thin coat of copper flux over it and fire. A curved piece must have a good rest on an asbestos form.

Now lay the perforated base for the *plique-à-jour* on this fluxed and fired copper and fire again until the plique-shape adheres as cloisonné wire would. This shape can be either copper, sterling, fine silver, or gold. Wetpack the colors tightly as you would with cloisonné, a mound on each cell. Dry the piece and fire very high; the enamel will seep to the middle of each cell if the fire is too low. You will have to repack once, maybe twice. But you may also succeed with just *one* firing, depending on the size of the cutouts. When all cells are well filled, stone while the copper is still backing the enameled piece. Sometimes, depending on the difference in expansion and contraction, the copper will peel right off and leave a clean back, with the coat of flux which has to be stoned down. Otherwise, the front and sides of the *plique-à-jour* have to be covered with asphaltum and the piece then immersed in nitric acid which will dissolve the tin backing of copper. The coat of flux protects the base metal of the *plique-à-jour* from damage. Since no soldering was involved, there are no discolorations or impurities, which solder causes so easily.

GRISAILLE

The white used in grisaille is an extremely fine-ground powder which can be applied to the dark enamel background in many ways. Grisaille white contains all the shades from opaque to the finest veil of white through which the background remains visible. It depends on the design which medium (oils or water) the craftsman chooses. Transposing a drawing which is normally done black on white into the opposite is quite tricky. For instance, when drawing an eye, one must draw the white of the eye with its curve and highlights, leaving the pupil as dark as the background permits. It is a good idea to get some experience first with white tempera on dark paper.

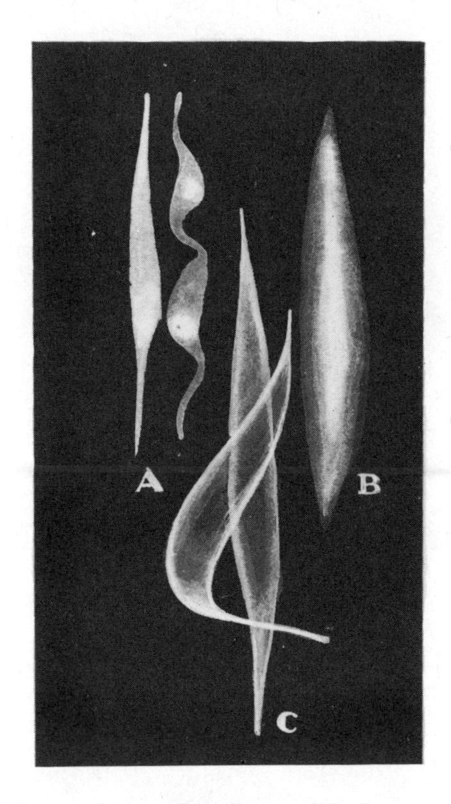

When making a grisaille enamel, the problem is to translate a black-on-white (or gray) drawing to a white-on-dark enamel.

The only source of supply for grisaille white I have found is Schauer, in Vienna, Austria. The number of the grisaille white: 497. To mix it with oils, prepare only the amount you will need in one session of work. On a small, rough pane of glass, mix with a horn spatula adding just a drop of *thick oil of turpentine* to make a thick paste. Now you have a choice of oils as thinner:

Oil-of-lavender lets the white flow from the solid white to fine, thin shadings which blend into the color of the background without a sharp, white contour.

Turpentine will keep a sharp, clear line. It doesn't flow as the paste of grisaille does when used with *oil of lavender* or *oil of cloves.*

Plain distilled water has yet another effect: it offers the whole range from opaque white to the finest, veil-like appearance, but each stroke of the paintbrush leaves a small, delicate white rim. It is not the right medium for figures or portraits. Naturally, when using water, you should not mix the powder with thick oil of turpentine, but with water only. While painting with oils should be completed within one or two firings, grisaille with water can be fired three times without losing its body.

Firing grisaille, which is done with oil, requires some experience. Dry the piece on top of the kiln until it is quite warm. Then, heat it in front of the open kiln: hold it on its trivet into the opening of the kiln and take it away again. Repeat this several times to increase the temperature and to help the oils evaporate; you will see the fine vapors. The grisaille begins to change to a yellow, and even a brown when too

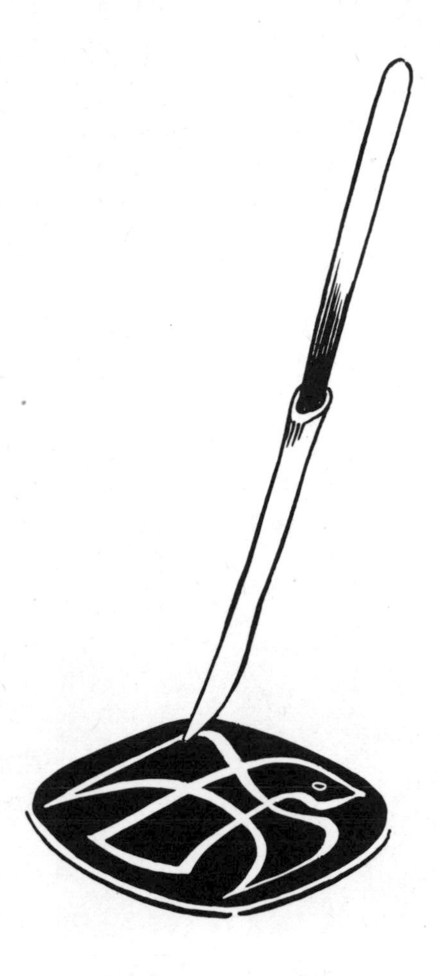

Tool made from a chicken bone.

much oil of turpentine has been used. The kiln has cooled by now and you may insert the piece for a moment; then take it out again quickly and test the grisaille at some less important point. Try to draw a line through the white with a well-pointed small stick of hard wood and see if the white is dry and powdery enough. As long as there is still too much oil the color will chip. If it has already fused, you no longer can draw a line through it. The trick is to catch the right moment. It is really not as difficult as it may seem. Now, the small repairs or refinements of the painted grisaille, like a negative sgraffito, can be done. Contours are sharpened, some color can be taken away completely. When this is done, dry the piece again and fire at about 1350° — dark red, until either the white has fused or it shows a slightly glossy surface.

In firing, the craftsman can decide *how intense* he wants the grisaille. The more heat, the more it will fade to the grays; if overfired, it disappears.

White should never be added after drying, only after firing. The tool to do this negative sgraffito needs some special mention. Metal would scratch and hardwood quickly becomes very soft and wide at the point. The ideal material, and one that is easy to find, is bone, a little hollow piece of chicken-bone. It is hard enough but doesn't scratch. It can be filed, shaped, smoothed with emery paper and it has a hole by which it can be fastened to a small stick such as the stem of an old paintbrush. When painting grisaille with water, the bone can be used right away after the painting has dried. White, mixed only with water, wipes off so easily that one has to be extremely careful. But add Klyrfire to the water and the grisaille white adheres better, can still be incised with fine lines, and fires with no problem. The firing is done at a temperature of about 1400°F.

Background for Grisaille

You may use opaques, but I prefer to use transparents which are good over copper without flux. They appear very dark but yet have a certain depth. Some enamels crack when refired; the craftsman may not even notice it for they "heal" again completely. Such enamels are not right for grisaille, as they will destroy the drawing. I have found dark blues to be reliable, and dark smoky gray which in two coats fires to a deep black. It served as a background for the miniature portrait on page 73. A transparent brilliant orange-red fired to a very warm brown, a special brown, in two coats over copper is the background for the small portrait on page 8.

While it is advisable to have the metal base slightly domed for wet-pack or cloisonné enamel, grisaille mixed with oil is safer on a flat or even slightly concave surface. The reason is that enamel will seep to the sides when fired on a domed shape; it would take portions of the grisaille with it and might cause cracks in the piece.

Copper is much easier to work with than silver. A rather complex design which had to be executed on fine silver gave me a lot of trouble until I built a small dike of enamel all around the plaque to counteract the sliding of the enamel and the expansion of the silver. This helped and the grisaille fired very well. The little wall disappeared later under the bezel and for the most part had already blended into the enamel anyway.

Pegasus on the Pasture. Grisaille enamel. (Owned by Pastor Martin Niemoeller)

Life drawing for grisaille medallion.

Medallion in grisaille enamel with fine gold lettering.

The Perfect Surface

The enameled background of a small and precious grisaille should have the perfection of an optical lens.

To achieve this, fire two or three thin layers of immaculately clean transparent over the metal. Then stone the surface with the same care you would take with gold cloisonné and brush under running water until no impurity is left. If there are some small white spots in the fired enamel, the color was not freshly ground or re-ground (if you used older enamel), nor was it washed well enough. The little white spots can be removed with a diamond-impregnated point in a flexible shaft. Wash the piece again, fill and refire. After each coat is fired you should check the enamel for such flaws. Actually, the craftsman should be looking for them while wetpacking the enamel. They are visible prior to the firing and can be removed with small tweezers.

When the surface is perfect after stoning, rest the piece upside down on the trivet. Of course, the plaque has a metal rim which touches the metal of the trivet and prevents the enamel from getting stuck to the trivet or from showing marks.

Fire fast and high – you will enjoy the result! (The same method applies to fine cloisonnés which are flash-fired.)

If grisaille is used on larger pieces, two clean layers of enamel for the background, sifted very carefully, are sufficient. The strokes of the brush should have the ease of Japanese painting. A little more pressure will spread the hair of the sable brush and widen the painted line to a shape. A small drop of color will flow and blend into such a shape quite naturally if the artist knows how to apply the technique. Try it.

Last Touches With Fine Gold or Fine Silver

Good grisaille enamels are precious enough to have their finishing touches done in fine gold or fine silver. No luster, please.

The gold is a brown powder and is called "sponge" by the refiner and "Roman gold" by china-painters. I recommend the sponge, which is packaged as a dry powder. It might be necessary to grind it to a maximum fineness between two pieces of rough glass. Like grisaille, the powder is then mixed with *thick oil of turpentine* except that "gold facilitator" is used as a thinner. (This may be obtained from art supply stores which sell china-paints.)

Gold is applied in a good, solid manner (unlike grisaille which is intended to flow; gold would be very ugly then and would burn away). When the gold is dry, you can scratch it with the chicken-bone. (I am sure you will learn to treasure this tool as I do.) Like grisaille, the gold is fired in moderately low heat. After it has fused to the enamel, the gold still looks like mustard. But brush it under running water with the glass brush and you end up with a lovely, matte fine gold. Burnish it with soapy water, always in the same direction, and it is polished. After burnishing, I would still use the glass brush very gently to draw the strokes of the burnisher together. To check if the gold has fused, try to remove a little on a corner, where it will not show. If some comes off, fire again.

Silver is treated in the same manner.

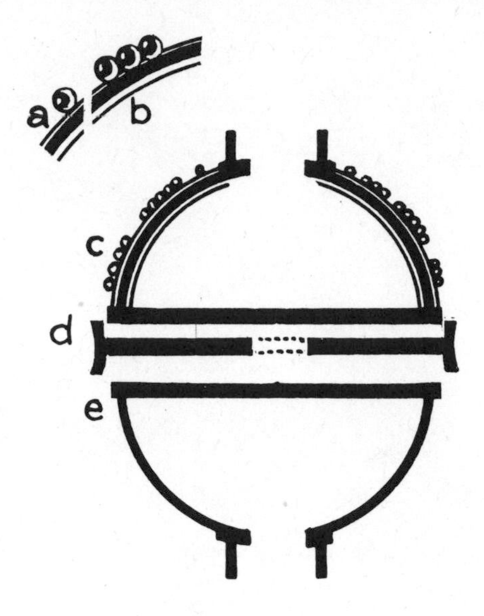

Detail of enameled sphere of gold Kiddush cup with grain enamel.

GRAIN ENAMEL

Start with "the perfect surface" described above as background for small, jewel-like grisaille enamels. This time either gold or fine silver would be nice as the base under a brilliant transparent color.

Grain enamel I call the application of minute gold spheres of exactly the same size. The gold must be non-tarnishing, 18 carat or fine gold. The small golden spheres are mounted in the thinly enameled surface; if the enamel were thick, the gold balls would "drown." The gold grains are glued to the surface with Klyrfire in a pattern which can be very contemporary or reminiscent of ancient gold granulation. Each small sphere is separately dipped into Klyrfire and set into place. I like to work under a hot light which dries the Klyrfire rapidly and thus keeps the spheres in their places.

When firing, hold your breath and don't let anybody talk to you. Again the "surprise" technique: fire fast and high. The enamel will be hot and sticky before the spheres have a chance to roll off, and *they will not,* as shown in the enameled sphere of the golden kiddush-cup on page 104.

Gold grains which did not remain in place in a design with grain enamel can be pushed into position while the enamel is glowing-hot. A small depression may remain and this should be filled with enamel.

Fine gold grains are fired on to a thin, even coat of transparent enamel over precious metal. The effect is comparable to gold granulation. Size 1⅞ inches.

Pewter box with enamel of gold granules in handle. Box by Frances Felten. (First award, Society of Connecticut Craftsmen, Annual, 1965)

How To Make Small Gold Spheres (for grain enamel)

To make a limited number of gold balls, use all your small leftovers from bending cloisonné wire. As far as possible, cut them to equal lengths and heat them on a block of charcoal with a blowtorch until they melt and turn into spheres. To separate the different sizes, use sieves of different mesh.

If a large quantity is needed, cut very thin fine-gold sheet or 18 carat, non-tarnishing gold as you would cut solder. Try to get the chips as similar as possible. Fill a layer of ground charcoal into a small graphite crucible or a box folded from a piece of stainless steel. Add a thin layer of gold chips and cover these again with a layer of charcoal, alternating layers of charcoal and gold chips until all the gold is between charcoal. If you can help it at all, don't let the chips touch one another.

Heat this container in a kiln capable of firing as high as 2000°F. or more. Then let it cool and empty the container into a pan filled with water. The light charcoal floats while the gold which has fused into spheres settles to the bottom. Gold balls and charcoal are easy to separate.

Use different grades of sieves and keep each size of balls in a small glass bottle.

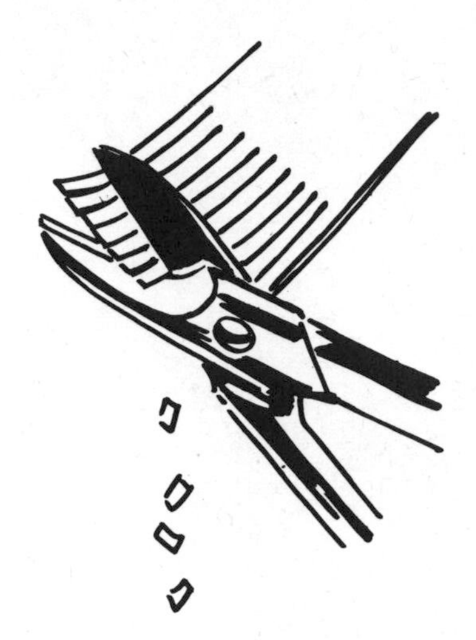

Cutting wire for small gold balls.

Melting the small gold chips into balls on a charcoal block.

Separating sizes of gold balls.

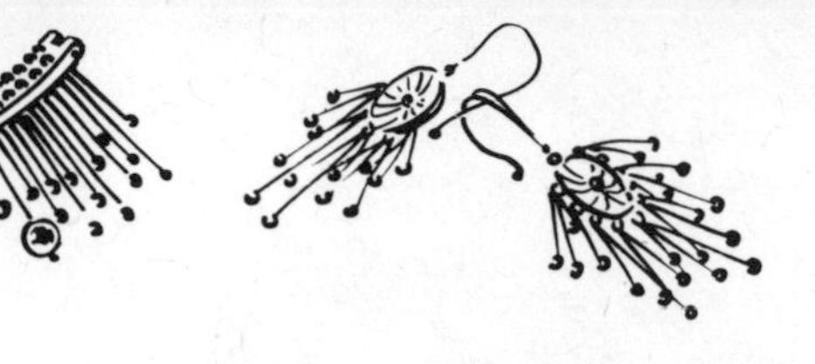

Jewelry made with tiny gold balls.

A Playful Way To Enamel Tiny Gold Balls

You will need: a blowtorch, mouth-operated; thin round gold (or silver) wire, 30 gauge; a spoonful of *finely* ground clear translucent enamel – dry; Fluoron solder flux; and a glass of cold water.

The whole procedure can be done on your metalsmithing table, or you could prepare a large number of the small pieces with enamel, hang them over a trivet, and fire them all at the same time in a kiln. I once had to repair an antique Indian bracelet with hundreds of those small gold balls enameled in emerald green and dangling in bunches from little rings. So I explored ways of doing it and finally adopted this method:

Dip the end of the gold wire into Fluoron. Heat this end with the blowtorch while holding the long piece of wire in your left hand. Let the end melt into a small ball, dip it into water to cool and then dip it into Klyrfire. Now dip only the ball-portion of the wire into enamel and then hold the end into the hottest part of the flame while you blow to provide plenty of heat. The enamel melts right over the entire surface of the little sphere. Cut off the length you want and then make another of these balls-on-a-wire. It is really fun!
Then refire, fast and high.

Fine silver can be used, but the different expansions of enamel and silver can cause trouble and cracking if the wire is rather thick. Since the charm of this technique is in the minuteness, I strongly recommend using gold.

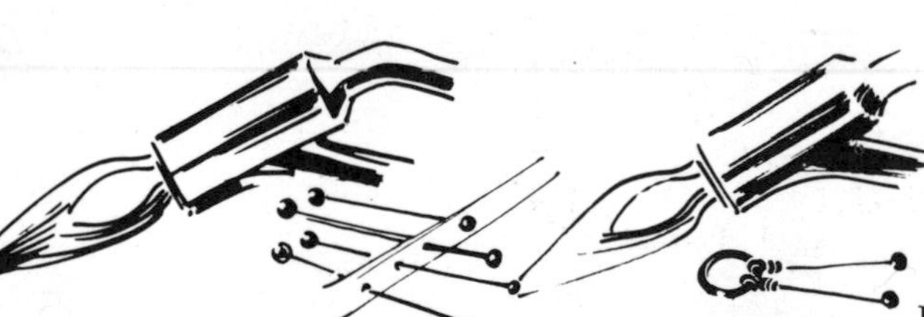

Enameling the gold balls.

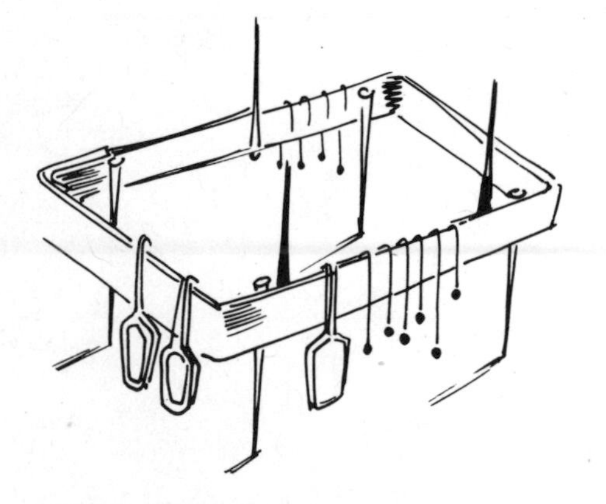

Trivet for firing very small enamels.

MERCURY GILDING

Mercury gilding has never been surpassed by modern electroplating. The quality and lasting brilliance of medieval pieces has not tarnished in a thousand years. Some frequently used areas may be worn a little, but on the whole, it lasts!

True, the craftsman has to know how dangerous it is to work with mercury and how deadly the fumes can be when they are inhaled. I advise: Wear rubber gloves and work with the wind outdoors or under a strong exhaust fan. No craftsman should ever let somebody else do this kind of work, and only the sincere and reliable artist should gild in this manner. He will achieve a warmth of texture and color that no other process can give.

One more advantage of mercury gilding is that on the same piece certain areas can be gilded while others remain the original material. It is almost possible to paint intricate designs with gold. This technique can produce three different hues of metal when working on copper and sterling. Both can be partially oxidized with liver of sulfur as well.

The work itself is rather simple and the necessary equipment as modest as one would expect an old technique to require.

Mercury readily alloys with gold and can again be separated by heating the gilded area to cause the mercury to evaporate.

1. To prepare the areas to be gilded we have to treat them first with a liquid made up of approximately 4 grams of mercury metal dissolved in a small amount of nitric acid, about a tablespoonful. This is then diluted with one quart of water. Keep the solution in a bottle with a glass stopper, or in some other well-sealed container.

2. Coat the inside of a chamotte (firebrick) crucible with clay or ochre. Dry and heat it until it is glowing red. Place into the hot crucible 1 part fine gold, rolled very thin and folded over, plus 6 parts mercury.

Cover the crucible and hold it with tongs, shaking it until the gold has been dissolved by the mercury and a white, pasty substance has formed. Pour the contents of the crucible into cold water after the crucible has cooled. This white paste must be kneaded thoroughly with a wooden spatula and should not be very moist. To remove the liquid you can squeeze the paste in a small suede bag. You now have gold-amalgam. Store it under water in a bottle with a glass stopper.

3. *The process of gilding:* The piece to be gilded must be very clean and free of grease. The areas to be gilded are painted with the liquid we prepared first (mercury-nitric acid and water). Rinse under water. A thin white layer of mercury will remain visible. If this layer is not *white*, the metal was not clean. Use a copper spatula to take some amalgam from its container. Take as much as this piece will require, place it on the area to be gilded, and spread it with a moist brass brush, copper spatula, or wooden point, as evenly as possible.

To cause the mercury to evaporate, warm the piece very slowly *outdoors* with your back to the wind so that fumes will be blown away from you. You can use a "bernzomatic" torch, or even charcoal. The important thing is to warm it *very, very slowly.* The heat is applied from below. Gradually, the places covered with amalgam begin to shine like a mirror. Now use cotton-covered wooden sticks or "Q-tips" to spread the mercury. As the piece becomes warmer, the mercury evaporates in white and extremely poisonous fumes. The surface begins to look dry and gradually the color changes from white to matte gold. The process is finished.

Either brush with soapy water and a brass brush to a matte finish, or burnish with soapy water, always using the burnisher in the same direction. The gilded surface is not completely smooth and the burnished grains will have a more intense polish than the lower parts. This is the charm of mercury gilding and an indication of fine handcraft.

For enameled pieces, mercury gilding is less dangerous than electroplating with its strong current.

6.

Making Finished Objects

THE NORWICH CROSS

Enameling

The 15-inch disk in silver cloisonné for the middle of the Norwich Cross was made first.

The enamel has been executed on 16 gauge copper, which is domed for better reflection, for easier stoning, and because only a domed piece of this size will hold its shape.

The large plaque was bright-dipped (immersed in a bath of one part each sulfuric acid, nitric acid, and water for 5-30 seconds) and the design incised with a sharp point. Then the plaque received a coat of copperflux and yellow transparent, mixed in equal parts, and sifted over Klyrfire. Firing flux and color at the same time kept the first coat rather thin and still provided the basic color, which would blend harmoniously with all other shades to come. The incised design remains visible under transparents. Counterenamel was sifted and fired in a good coat and again fired. Starting at the center, the symbol was applied and held with Klyrfire in the manner of mosaic, each small rectangle of thin 18 carat N-T gold is part of the radiating design. The cloisonné wires were placed at the same time, taking care that no wire rested on the gold, only on the enamel. The wires had to sink into the enamel during the third fire. Two different sizes of wire have been used: 0.8 mm by 0.8 mm for the contour of the symbol "PX," and 0.8 mm by 0.30 mm for the "rays."

Cross for Church of the Resurrection, Norwich, Connecticut. Height: 8 feet. Wood by Kenneth Lundy; pewter by Frances Felten.

The firings all had to be done in a large kiln, which was kindly provided by Bovano in Cheshire, Conn., about forty miles away from my shop. To overcome the danger of distortion while transporting the enamel with wires and gold only glued to the first coat, I added small amounts of enamel to all strategic points and soaked them with Klyrfire. It worked.

The trivet for such a large piece, which has to fit the work of other craftsmen, is of major importance. It must give a perfectly flat rest to the rim of the plaque to avoid losing the shape. Also, the kiln has to be big enough. If the edge of the plaque gets too close to the heating elements, the enamel around the rim will not only fuse, but the silver cloisonné might burn, while the enamel in the middle of the plaque barely melts. It would be wiser to have silver cloisonné only about 3 inches inside the edge. Copper or gold wires create no problems of this kind.

Important! Before inserting the plaque into the kiln, several samples with the same metals and enamels must be made and timed exactly while in the kiln. They should be placed where the edge will be AND where the middle will be. This shows the limit of firing time required for the large plaque.

Opposite page:
The enamel for the Norwich cross. Silver cloisonné with gold mosaic. Diameter: 16 inches.

The third firing had fixed the silver cloisons and the sheet gold to the plaque and I could now wetpack and fire, wetpack and fire again, until all cells seemed to be well filled and the shading was right. Some enclosed small areas had been filled with a rather sharp transparent turquoise to give more life to the ambers, yellows, greenish grays, and browns. Seen from a distance, this turquoise blends beautifully into the whole, and, when one looks at the plaque closely, these areas enhance the enamel like precious stones.

The stoning could be done in my own studio. Of course there were still some low areas, and small changes and repairs were necessary. When all wires were exposed, smoothed, and polished, and the plaque was immaculately clean, I added the last wetpacking at the plant. To keep the shape perfect in this last fire was most important. I had two long iron forks handy to press down what might not be in proper shape, while the plaque was still glowing red. Later on, when the wood, pewter, and enamel were assembled it was truly a great pleasure to see how well things worked out.

The Wood

Executed and described by Kenneth Lundy.

In planning the structure of the cross frame, several factors were considered. It appeared that the wood selected must be appropriate for its function, thoroughly seasoned, hard for the sake of durability, and must take a finish complementing the enamel and pewter. The final selection of wood was straight grain, selected mahogany of matching pieces.

A number of features were considered during the developmental stage, such as rigidity of the finished unit, hidden joints where possible, the problem of tapered planes and angles, the securing of the heavy pewter ring holding the enamel, the securing of the pewter strips, and the provisions for hanging the finished cross. With this focus in mind, a detailed scaled drawing was made including all views and some detailed sections of the cross. It cannot be emphasized strongly enough that the planning stage is the most important step, thereby leaving little to happenstance during the actual construction. It is true that minor steps can be overlooked while planning, but this process reduces modification or changes to a minimum.

The wood used in the cross was dressed one inch. The back of the cross was constructed first with the arms intersecting in a flush, half-lap joint. The edges of these pieces were then beveled to accommodate the angle of the arms of the cross. The next step in construction was to make the front and side pieces of each arm. The joining of these pieces was critical as this had to include hidden joints, proper bevels, and a strong construction. The idea of a miter joint and splint was considered in planning but was not feasible for six-foot lengths. A modified half-lap joint with tapered edges was made, giving a rather large, strong, glue area. Kerfs were then cut in the face of the arms to accommodate the pewter square wire and the same kerf was made on the reverse side of the wood as a point of reference. Before final assembly and gluing, the circumference arcs at the center were carefully laid out and cut so that the round pewter ring could fit exactly in place.

This detail shows the joining of one of the arms of the cross so that no seam can be seen from the outside. The pewter is inlaid into a groove; a similar groove is cut on the inside to receive the screws which hold the pewter inlay in place. You can see one of the holes in the wooden backing made to admit the "spring-grip" screw driver.

Detail of recess which holds the assembly of the enamel plaque, plywood backing, and pewter framing.

Back of the cross. Screws were inserted into the pewter-inlay rod from the back (see page 46). After the pewter inlay was secured, the screw holes were plugged with disks of matching mahogany.

Following this, a circular plate slightly larger than the pewter ring was turned and fastened with screws and glued to the rear of the cross. This step gave the necessary rigidity to the entire piece and at the same time gave backing for the pie-shaped pieces that were needed to raise the ring to the desired height.

The wooden portion of the cross was now ready for the insertion of the pewter wire in the kerfs and the placement of the ring. Suitable holes had been previously drilled in the back of the cross and it was now a matter of reaching through with a screw-holding screwdriver to secure the pewter wire and ring with self-tapping metal screws. Tapered wooden plugs were then inserted and flush-mounted in these holes. This gave the back of the cross a finished appearance. Suitable eyebolts were now secured at three determined points on the back for the purpose of mounting.

The last step before completion was the finishing of the cross. Lampblack mixed with a vehicle of linseed oil and turpentine was used. Shading of the stain was accomplished by fine steel wool and several coats of hard wax were applied for the final finish and delivery.

The Pewter

Executed and described by Frances Felten.

In setting the enamel plaque for the cross the first consideration was an accurately dimensioned working drawing. A disk of $\frac{3}{4}$ inch thick plywood was cut to fit the inside diameter of the outer pewter rim, which is 12 gauge thick. This plywood serves as backing for the enamel plaque and as a spacer that holds the frame in complete juxtaposition with each arm of the cross. It also stabilizes the round shape of the frame.

The plywood was firmly centered in the recessed space of the wooden cross and fastened with brass screws.

A plain cylindrical band of 4 inch pewter, 12 gauge, was fitted around the plywood disk. It made firm contact with each arm of the cross. This band was then soldered. A $\frac{3}{16}$ inch square pewter wire was soldered inside the top of this pewter frame.

As the next step, a conical development was made as a pattern for the inner part of the frame. This conical section was fitted in contact with the enamel. The square wire was beveled by filing to the same angle as the conical band would produce. Then this inside band was marked where it had to be cut and soldered.

After the solder-joint was cleaned, this conical band was pressed into place. To hold it, a second pair of hands was needed. It was soldered to the square wire and outer rim while the plaque was placed in the cross. The photograph shows that a small ledge is advantageous to hold the solder. When soldered, the whole frame could be taken out and the top leveled.

The next step was to *tack* the four supports, which flow from the pewter rim into the four inlaid bars of pewter. These bars extend over the full length and the ends of the arms of the cross. This tacking had again to be done while the enamel was placed in the right position in the center of the wooden cross. The fitting of the metal to the wood was a very precise operation. The soldering could then be done with the pewter frame separated from the wood.

The outer rim is fitted.

1. The four supports are polished and ready to solder to the outer rim. First, each is tacked into place on the cross. Then the whole rim is taken off.

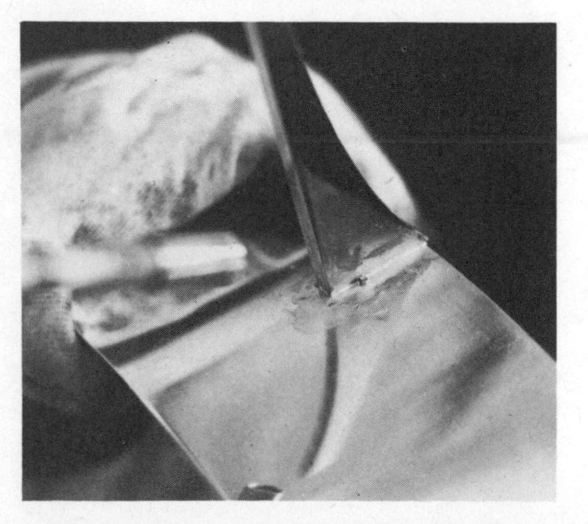

The bar wires fit into grooves 1/16 of an inch deep. The bars, after being shaped to the four ends of the cross at the same obtuse angle of the wood, were left longer than necessary. After the fitting was done within the groove, a board, 3/8 of an inch thick, was covered with asbestos sheet. A wooden channel, 1/4 inch wide, was screwed to the asbestos-covered board. Where the soldering had to be done, a section of this channel was cut away to avoid burning the wood. Inside this channel the bars would meet and be kept perfectly straight.

During the soldering operation the whole pewter rim had to be raised exactly as high as the thickness of the board and asbestos sheet combined.

Before soldering, a piece of wax paper was placed under the board, since the wood of the cross had already been stained and completely finished.

Now the last soldering could be done. The joints were handfiled and finished and the bars dropped into place. Screwing the bars into place from the back was the last step to complete this work.

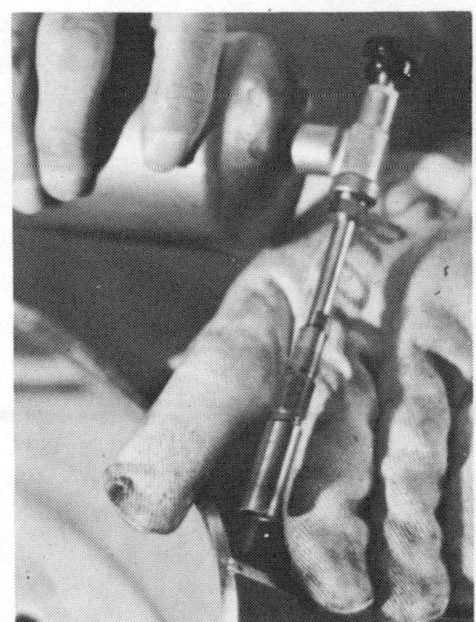

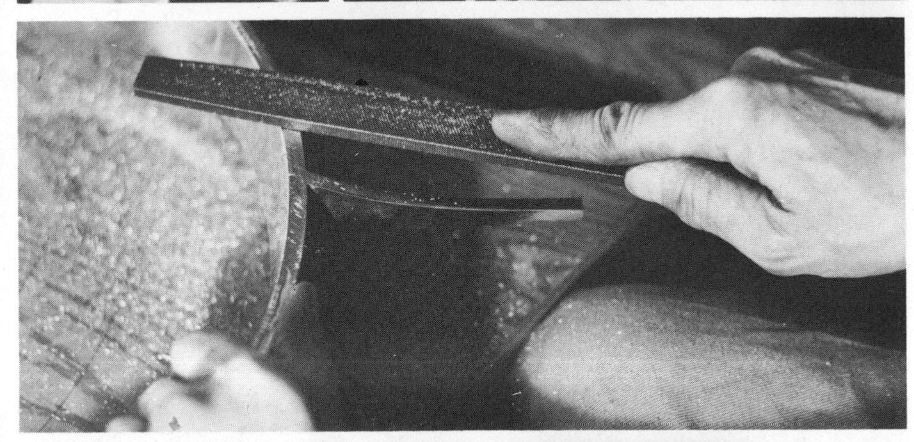

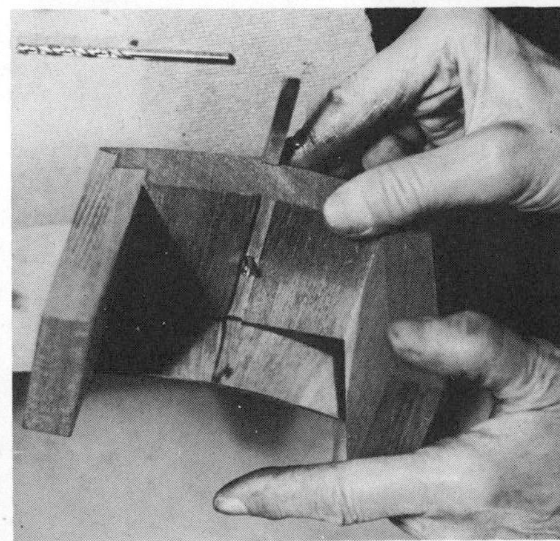

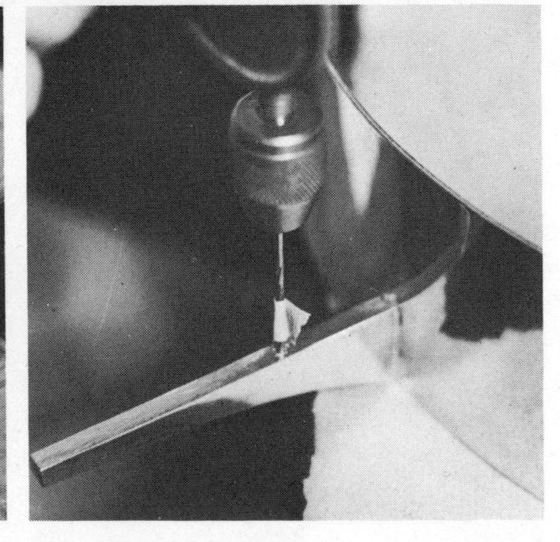

2. Each support is well soldered, small parts of it at a time, to hold securely.
3. The inner, conical ring is soldered. The cross is completely assembled, the enamel in place. The conical ring, still open, is tacked at short intervals. A second pair of hands is needed to hold it down closely over the enamel, permitting no gaps, and fitting to the enamel perfectly. The rim is then removed again and soldered well.
4. The face of the rim is leveled and work on the inlaid pewter bars can begin.
5. This shows a sample cross-section of the wooden arms of the cross. A hole is first drilled into this sample to measure the exact length of the screw. (On the cross, these holes are pre-drilled.) The next step is to drill holes into the groove of the four arms. To locate the place on the metal bar where the screw will cut its own thread, a sharp, straight point driven from underneath was used.
6. Tape is wrapped around the drill, which has the root diameter of the screw thread. Tape stops the drilling when necessary depth is reached. The screws will cut their own thread when the cross is assembled. The screwdriver used must *hold* the screw head, since the back of the cross is enclosed and has holes on the right spots. These holes will be closed with small wooden circles when the assembling is done.

7. The completed middle section is buffed.

8. The whole middle part is lifted ⅜ of an inch so that the inlaid bars and the supports will meet . . .

9. . . . when the asbestos-covered board, ⅜ of an inch thick, is placed over a piece of wax paper.

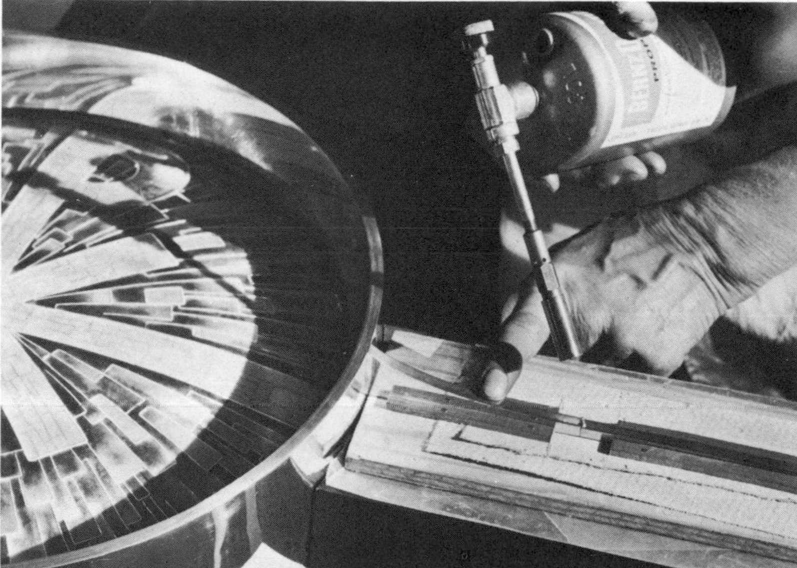

10. The wood has been stained and completely finished before this last soldering operation takes place.

11. The filing and hand-finishing of the joints.

Bishop's ring. Madonna and Christ Child flanked by symbolic black and white infants.

ENAMELED BISHOP'S RING

The construction of the metal part has been explained on page 40. Here is the enameling procedure:

1. The first coat of inside enamel is sifted over Klyrfire. To reach every space, a very small sieve (the homemade kind) was used. The tip of the little finger was helpful in tapping the enamel tightly along the two rims.
2. The design of bent gold cloison wires was set right onto the engraved gold on top. A first coat of wetpacked enamel held it in place during the *first fire*.
3. After pickling and careful washing and cleaning in ammonia and under running water (to remove every possible trace of acid), the top received a second coat of enamel, held in place with drops of Klyrfire.

Now the lettering around the top was placed and imbedded into a tightly-packed first coat of enamel dark above the letters, close to the rim, light in the band of letters, and dark again underneath the lettering. The same dark enamel was applied around both sides of the shank, to prevent oxidation of the solder-joints. All of the enamel was held in place with drops of Klyrfire on top of the wetpacked enamel. No mixing of Klyrfire with enamel; it does not fire right. I placed the ring very gently on a shiny surface, when I did not hold it in my left hand while working on it. The front was not yet fired and had to stay intact. Except for the space left for lettering around the shank, the whole ring was now covered with enamel and could be *fired for the second time.*

One letter had slipped. I ground it out with a diamond point, cleaned the place carefully, and sifted first a second coat of enamel over Klyrfire over the inside. Some lettering to go inside was gently pressed into this dry enamel and then a light mist of Klyrfire was sprayed over the whole inside.

The ring was now ready to receive cloisons along the shank, first on the side where the letter had to be replaced at the same time, again tightly packed, and held with drops of Klyrfire. There is a danger that letters and enamel will slide if the piece is too moist. Kleenex soaks all superfluous water out before one proceeds to the next section. After drying, *the third fire.* The same procedure on the other side of the shank and the next, *fourth fire.* Now back to the first half, which got its second coat, and the ring its *fifth fire*, and the second half, plus wherever some enamel might be missing, *sixth fire.* Meanwhile, the inside evened out beautifully, and since all went well, the ring was ready for the finishing: the usual procedure of stoning with medium and fine carborundum, with scotchstone and wet-and-dry-emery paper. Finishing and polishing of all metal parts was done by hand since I would not immerse the ring again into acid. Now there was the choice between a fast and high flash-fire, which would produce brilliance but might get scratched in years of use, or hours of patiently rubbing the piece with cerium oxide and wet felt. A motor cannot do it, it has to be done by hand. I decided on this latter way because of the touch and quality only a hand finish brings out.

Variation of the bishop's ring. The theme of the ring is stated in gold cloisonné lettering around the top and the shank: "God created man in his own image, in the image of God created He him."

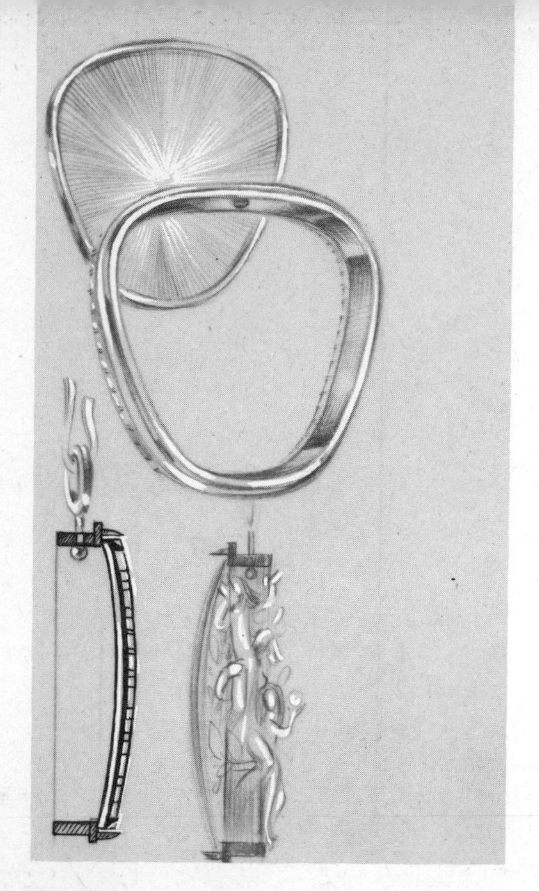

PIN-PENDANT WITH CARVED GOLD FIGURES AND CLOISONNÉ

This lighthearted little pin-pendant was a Mother's Day present especially designed for the woman upon whom it was bestowed. Whether worn as a pin or on a chain, it turns easily so that either side may be seen. One might say it is four pieces of jewelry in one.

The metal chosen for enameling is 18 carat, non-tarnishing gold, engraved on both sides for intense reflection through the transparent enamels and for a better grip of the enamel on the metal. One side is executed in gold cloisonné, including the first initial of the owner. The reverse of this is enameled in clear emerald green and provides a background for the small gold figures.

When the enamel was completed and in no more danger of changing size or shape, I made the bezel-frame of 18 carat gold. This did not have to be non-tarnishing so I could use a harder alloy.

The small figures were made from heavy gold wire. Filing, carving, soldering the little sculptures, bending them into their graceful shapes with pliers required special care. To avoid damage to the figures, the nose of the pliers was covered with several layers of tape. When all parts were finished, polished, and fitted into position, I soldered them piece by piece into their "stage," keeping in mind that they must remain "easy," seeming almost to float over the precious green enamel.

All solder-joints were covered with ochre during the many separate soldering operations. Then, a last touch with a soft buff, and enamel could be set.

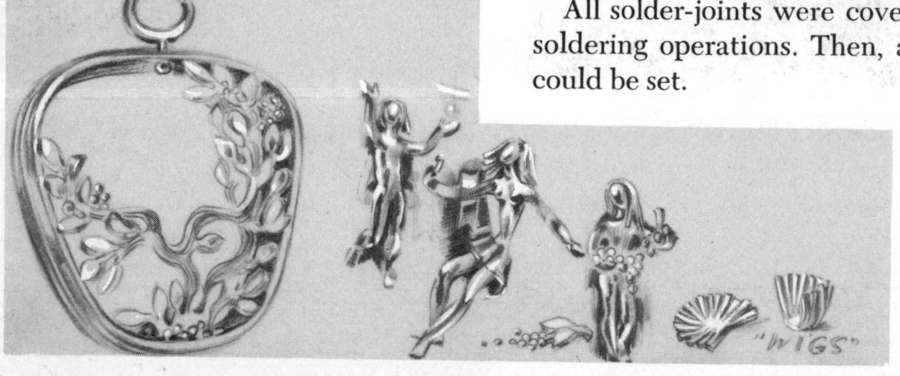

Front and back of pendant with gold figures sculptured *en ronde* on enameled background. The small figures are filed from gold wire.

Opposite page
Cross in pewter with champlevé and cloisonné enamel, detail. Height (full piece): 13 inches. Pewter executed by Frances Felten.

WHY ARE
YOU AFRAID
OH
MEN OF
LITTLE
FAITH ?
LOVE YOUR
ENEMIES + BLESS
THOSE WHO CURSE
YOU + JUDGE NOT
+ CONDEMN NOT +
FORGIVE AND YOU
WILL BE FORGIVEN

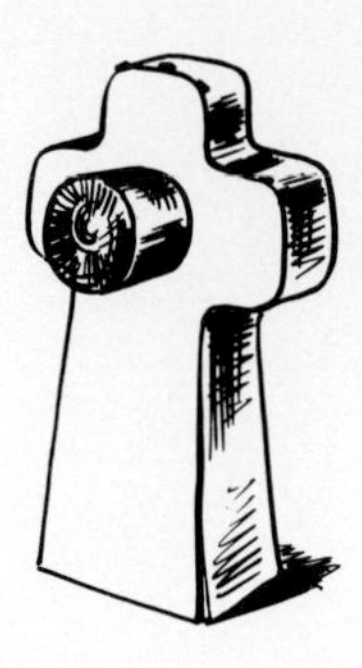

Wood cylinder to adjust front and back to right angle.

Solder flat, on asbestos. Hold with wire over asbestos sheet.

Fit pewter cylinder.

"Spacer" inside cylinder as backing for enamels.

PEWTER CROSS WITH CHAMPLEVÉ AND CLOISONNÉ ENAMELS

The front of this cross shown on the preceding page, has a gilded champlevé enamel with some cloisonné wires as a center, set deep into the pewter cross. Copper and silver cloisonné wire are the metals used for the center enamel. Its surface is flush with the gilded metal and stoned smoothly. A coat of wax brings the enamel colors to a mild shine. A silver cloisonné disk of the same size is set in the same manner into the opposite side.

Frances Felten executed the pewter and here is her description of how it is done:

Front and back are sawn out from 12-gauge pewter; both are buffed and polished. The front has been etched with lettering.

A good work-drawing should be made for anything as permanent as metal. From this drawing the pattern for the sidestrip is developed, excluding the bottom. This long sidestrip, bent to the shape of the cross, is soldered to one side.

A wooden cylinder, turned on the lathe, fits exactly into the holes of the cross and stabilizes it in the proper position. The second side of the cross is tacked with solder to the other part, while the wood is inserted. This careful soldering is done on top of the cross where it can be easily reached without moving the piece. While the wood cylinder is still inserted, the parts are tied together with steel wire, protecting the pewter with pieces of asbestos. Now the wood can be removed, after the cylinder is marked flush with the two faces of the cross. The piece is placed flat on a large asbestos sheet and soldered together. The marks on the wooden cylinder give the dimensions of a pewter cylinder to be inserted into the cross. Into its exact middle fits a wooden spacer, shaped on both sides in an angle corresponding to the profile of the cross.

The two plaques are cemented to this spacer and remain there while two rings, fitting inside the cylinder, are soldered in close contact with the plaques. These rings extend about $\frac{1}{16}$ of an inch over the face of the cross to make a ledge to hold the solder.

During these last procedures the two enamels are protected by two strong cardboard circles.

The rims of the cylinder can now be rounded and finished.

An ingot of pewter has been cast in plaster of Paris as a bottom weight. It fits into the foot of the cross. This ingot is soldered to the 10-gauge pewter base, the cross is placed over it, a line is drawn around the foot and, after removing the cross, the base is sawn out to fit inside the foot. The last step is to solder it flush with the bottom.

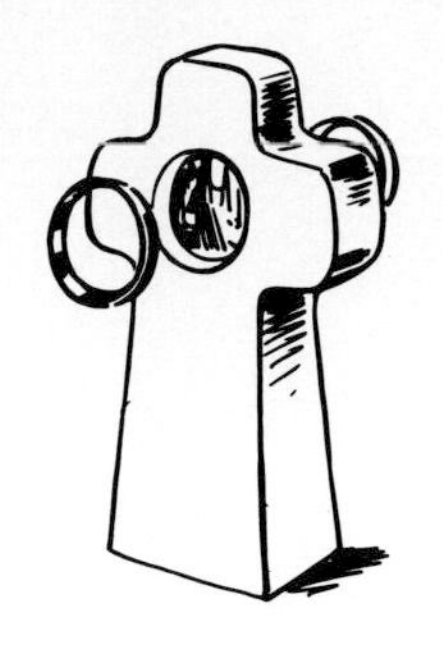

Enamels are glued to "spacer." Two pewter rings hold them in.

A SERENE CUP, GOLD CLOISONNE AND PEWTER

Cross-section with spacer.

No symbols of power, eternity, or love, no very valuable materials were used here, but abundance of joy and play, laughter and youth. This is what I wanted this cup to express.

Eighteen carat, non-tarnishing gold was used for all visible parts of the enamel — gold, not for its value, but for its ideal properties in cloisonné enameling: the color, the absence of discoloration in the fire, the great heat required. The cup and the metal foot are pewter. Modern pewter does not tarnish, contains no lead, and is nontoxic. It should be used rather thick and rests beautifully in the hand. Its color is warmer than that of silver.

The design shows three partitions which are soldered with gold to the metal base. They seem to be trees, but they started as wide bands of 18 carat gold (see sketch). Neat soldering, without gaps or surplus solder, was important. Enamel would certainly creep under every gap and cause cracks. Solder would cause black spots, pores, and impurities. The design itself is pure fun: eighteen happy little figures. The colors are soft: green, tan, off-white (achieved with an opaque white mixed with light gray and a coat of opal-white fired over it), an intense blue, little spots of red, some turquoise.

The construction is similar to that of other cups discussed in this book. All parts were made to fit and were assembled without glue, heat, or soldering. This permits dis-assembling, if necessary, without damage to any part.

Frances Felten did all the pewterwork and took care of the assembling.

Design for the bending of gold cloisonné wire for the foot of the cup.

Path of Destruction. Details of triptych. *Right:* Adam reaching for the apple from the Tree of Knowledge. *Opposite page:* Symbolizes the destruction of nature and civilization. Copper and copper cloisonné enamel.

TRIPTYCH: *PATH OF DESTRUCTION*

Left panel: Man reaching for the fruit of the tree of knowledge. Middle panel: Man stepping over his brother's body and asking, "Am I my brother's keeper?" Right panel: Complete destruction of nature and culture.

In this book we are primarily concerned with the techniques of making such a piece of enamel, but technique alone is not enough. In this piece the design and the thought behind it are of major importance. In my description the three are inevitably interwoven.

The plaques are copper and copper cloisonné enamel. The finished plaques are set between two pieces of dark wood, $\frac{3}{4}$ of an inch thick. The lower board is carved deeply enough to hold the enamels. Thick pewter (8 ga.) rims are fitted over the edges of each plaque and tightly into the cutouts of the upper board, almost like matting a drawing. Epoxy is used to hold all parts together.

The letters are of fine silver. Each letter has several $\frac{3}{8}$ inch long wire stems soldered to it. Small holes for these wires were drilled and each letter separately hammered into place. To avoid damage to the letters and the wood, a flat board was placed over the letters and with vertical, well-aimed strokes, they were driven into the wood almost flush with the surface. Before setting, each of the wire stems was dipped into epoxy.

A thorough description of enameling the human body follows. But here let me describe a few design problems presented by this triptych: How to express the abstract idea of the "path of destruction" with such finite materials as glass and metal? How to *show* the voice from above in my design? (The borderline between the serious and the ridiculous is hair-fine.) It could only be done as simply as craftsmen in medieval times might have done it. As I traversed a number of detours in my struggle, I realized that the artist's great danger is to be too symbolic. Enamel and metal are neither music nor poetry; they are terribly concrete. To simplify, to omit everything that is non-essential: this might be the path to true art and to truth.

How I wish that the painter-craftsmen reading this will go on from my experience. You should have many samples made for the colors you wish to achieve; try many different possibilities before you put the first *grain* of enamel on your piece. If it is to be cloisonné, you have already invested many hours of intense work that must not be lost. This getting acquainted with what colors can do gives you freedom when you are ready to start on your piece. Then forget about glass and wire: you must work out of your obsession with the theme. It is possible to make

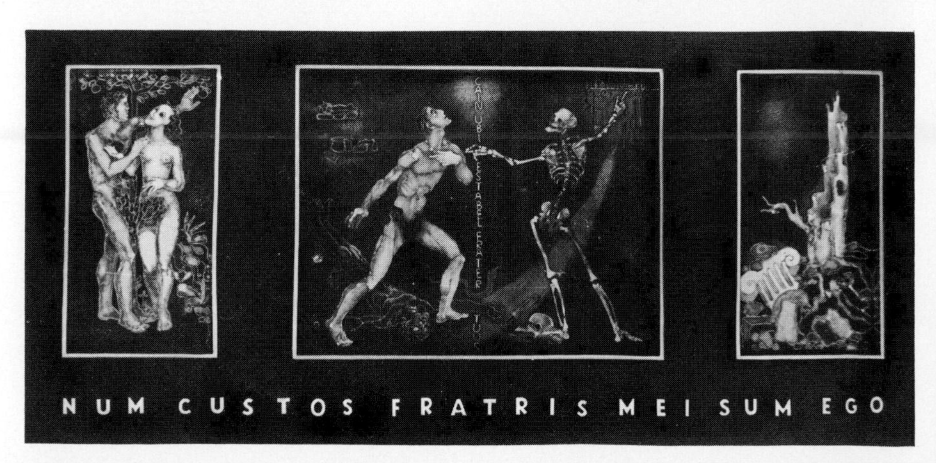

Path of Destruction. Triptych. Three copper cloisonné plaques in pewter frames set deep into a wood background. The lettering is in silver.

delightful enamels with the same approach as one makes delightful embroideries. But one can enamel with passion and say essential things with this medium.

When we discussed "Sifting" and "Stencils" earlier, I said that there are situations when a stencil, held high above the enamel plaque, is the best way to achieve a certain effect. The path in the middle panel of *Path of Destruction* was done in this manner, as was the light from which the lettering descends and the bit of light on the panel to the right.

To start with, for all three plaques I sifted the dark basic coats at the same time in order to retain their harmony and to make sure the different hues were placed right. In the righthand panel are some of those "textures" which the eye, but not the hand, can feel. There are opaque white and mean yellow spots like mould and decay beside marble-like off-whites and the grays of rocks. The marble effect is achieved in firing opaques, covering the basic dark transparents more or less thickly. These areas then are filled with opal white or a fine transparent gray, into which small chips of white or yellow opaque are sprinkled — not round granules, but sharp-edged little splinters of enamel. After firing and stoning you will have the visual effect desired.

Pages 92 and 93 show the outer panels enlarged slightly so that you may follow the lines of the wire with a pointer or pencil. See how these small pieces of metal strip, shapeless and meaningless by themselves, combine to produce both form and expression — often such strong expression that I wish I could paint like this.

Cross-section showing how enamel plaques, pewter frame, and silver lettering are mounted together in the wood.

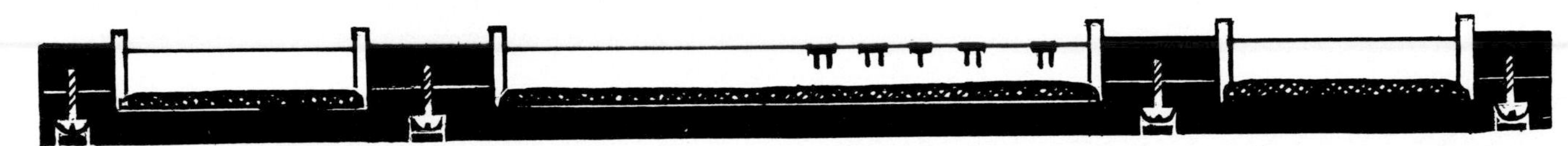

ACHIEVING A THREE-DIMENSIONAL EFFECT (THE HUMAN BODY)

Independent of what the metal background may be — whether it has a coat of flux or not, whether the first coat is a dark or a light transparent, two facts should be observed:

1. This first coat must be thin.
2. The cloisonné wires must be high enough to hold three to four more coats of enamel.

If the first coat is too thick, the wires will drown in it, leaving not enough room for further enameling. Or worse, they will sink in at a slant. When working with silver wire, avoid hard colors. The melting points of both are too close and, especially around the edges, some wires might burn off. A transparent background is advisable, to create space around the figures.

The safest way to gain experience in this technique is to draw the design with Underglaze *D* directly on the metal, fire a thin coat of flux or transparent over it, and set the wires, dipping them first into Klyrfire, as described before. The enlarged details on pages 92 and 93 show very clearly how small units of wire form the whole design.

Now choose the right opaque from the sample plaque (p. 27) which shows some varieties of skin shades you can enamel. With only this one

Silver chalice with gold cloisonné. Diamond set in platinum and gold rim and inserted in silver base. (Commissioned by Fairfield University, Fairfield, Connecticut)

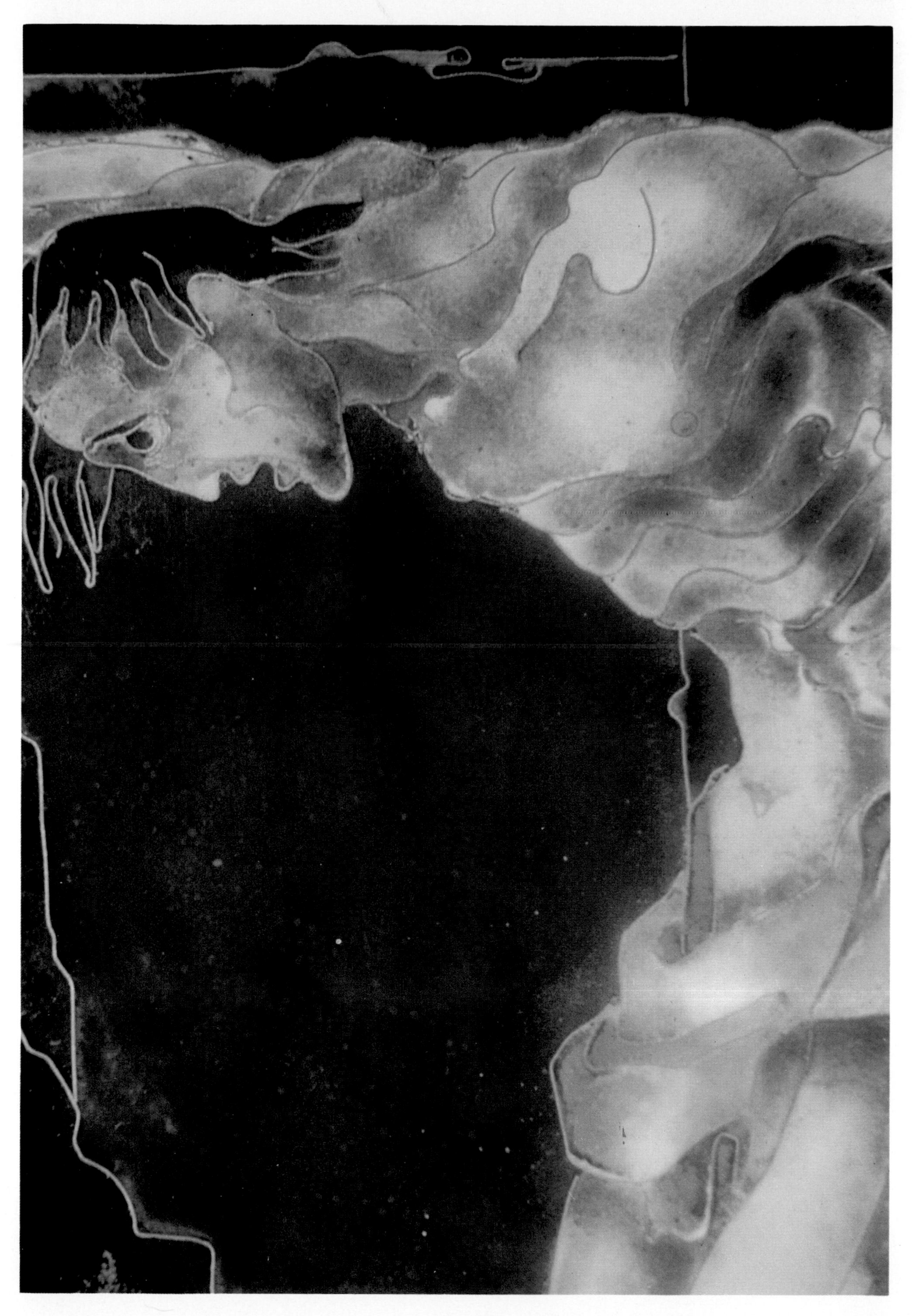

Detail of silver cloisonné panel (see sketch on page 102).

opaque build a kind of relief of the figure inside the cloisons. With spreader and a rather dry brush, model the foreground almost up to the rim of the wires, keeping in mind that this will create highlights in the finished piece. The hand at the very left and the shoulder of Peter (opposite) demonstrate this. The enamel should not be too wet if it is to stay in place. Klyrfire is not needed on flat plaques, but is needed on curved surfaces. Work over the whole plaque wherever this three-dimensional effect is desired. Fire the piece, watching very carefully NOT to overfire. Even in a normal fire the opaque will melt to a shiny even surface. To keep the sculptured effect, the enamel must just begin to fuse, producing the orange-peel surface. It is possible to soften the relief while firing.

When the plaque has cooled, check to be sure that the opaque has covered the whole figure. If there are still some bare spots, refill and refire. The enamel looks most unpleasant at this stage.

Next, shade with transparent. Your sample plaque will again tell you which hue. You may need a lighter and a slightly darker shade, but the relief will already provide for most of it. Go over the whole figure with these one or two transparents, not to fill the cells, but only to achieve the right hues. Fire again, also just fusing the transparent to the relief of opaque. The plaque will now look spotty and discouraging, but do not despair! Check again to see that the transparent covers the figure without leaving bare spots. A second application may be necessary. Do not overfire! Preserve the "relief."

(text continued on page 102)

Details from *Apathy* and *Betrayal* (see page 102) and *Violence* (opposite).

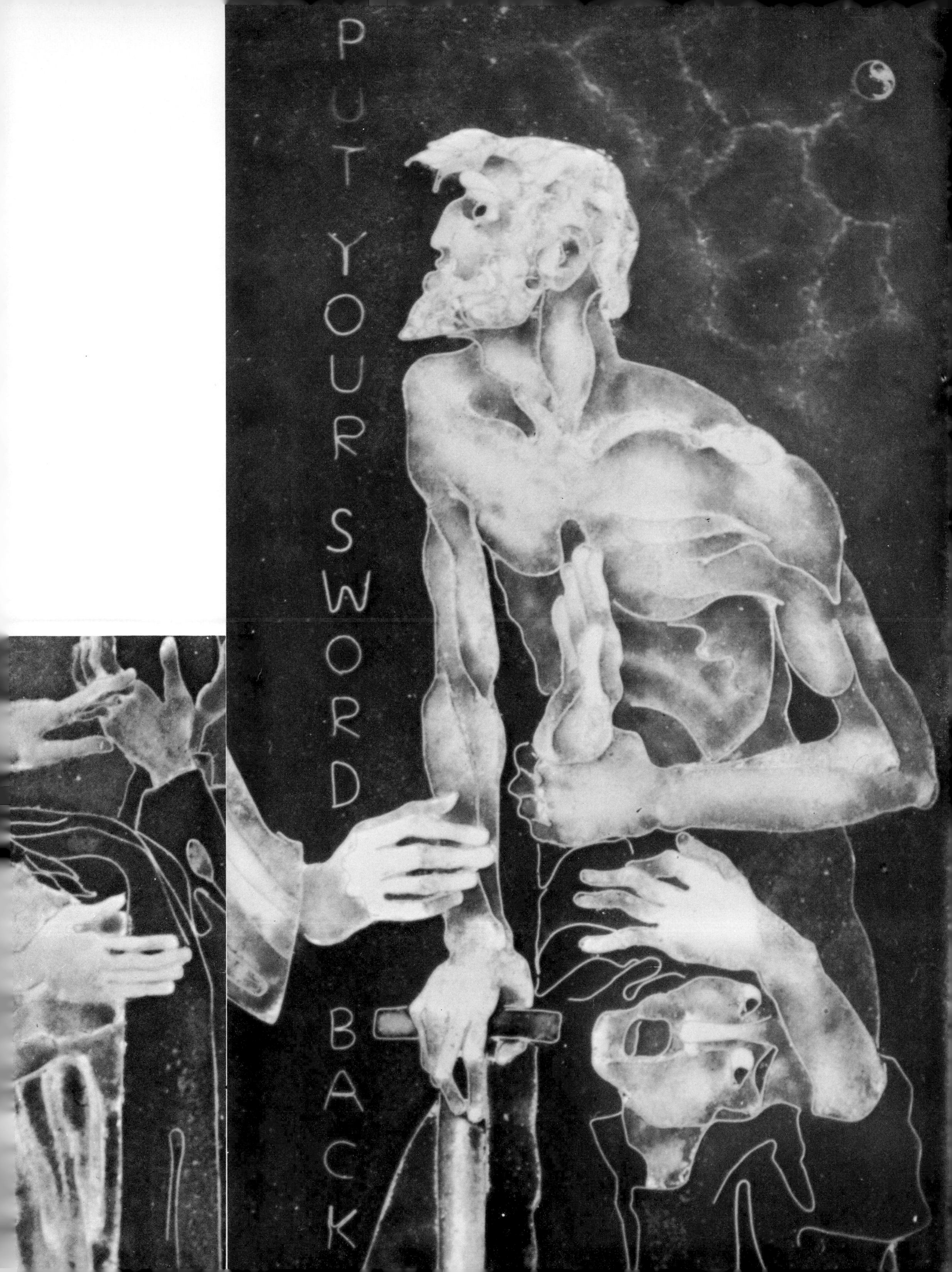
PUT YOUR SWORD BACK

Diana. Pendant of 18 carat gold with enamel on engraved gold background. Enlarged, original figure is 1 inch tall.

Fall in Connecticut. Detail of abstract plaque. Gold, silver, and copper cloisonné, with sheet gold fired under parts of the enamel.

Adam and Eve. Ring of 18 carat gold, enameled inside and outside.

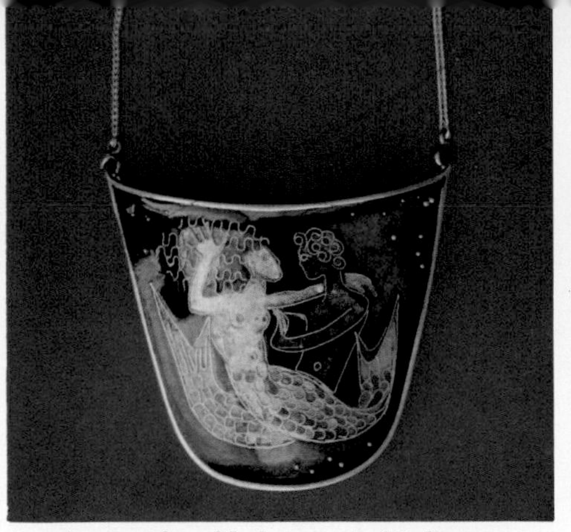

Pendant with gold cloisonné. (Collection of Stanley Lechzin)

Gold cloisonné box of silver, enameled inside and outside. Top plaque set in gold bezel with a string of pearls. (Collection of Peter Zeitner)

St. Eligius. Gold cloisonné to be set into the lid of a gold box. Diameter: 2 inches.

Three Graces. Grisaille enamel pendant, enameled on both sides. Height: 2 inches. (Collection of Hanne Zeitner)

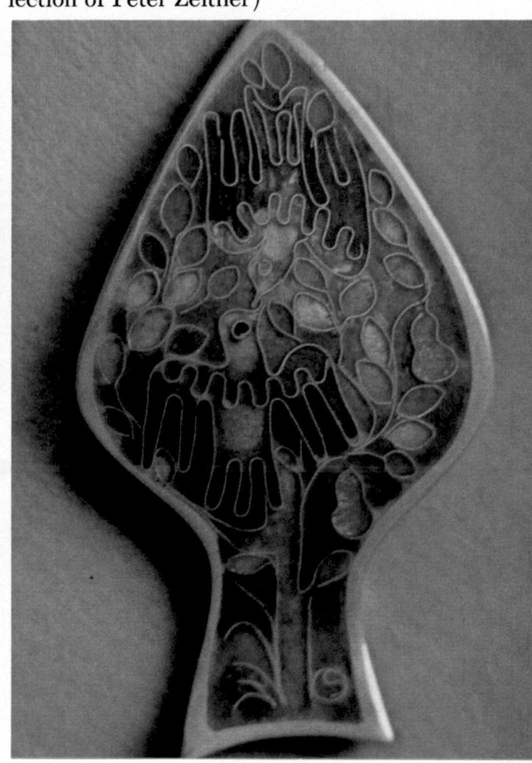

Gold cloisonné on cast sterling silver.

Pan and Syrinx. Gold cloisonné enamel, set in gold-plated, hinged silver box. Diameter: 3 inches.

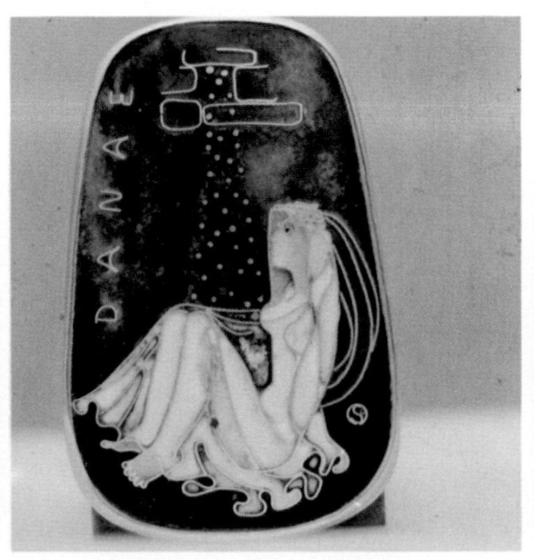

Danae. Silver box with enameled lid.

Silver cloisonné, set with silver and ivory.

(text continued from page 98)

The cells are not filled yet, except in some foreground areas. Now use flux (the same copperflux used for first coats) and fill the whole figure well up to the top of the cloisons. (If you planned to lay an even coat of transparent shading over the whole figure, this should be done and fired before the last coat of flux.) Pack the surface as evenly as possible; it saves a lot of stoning. Dry the plaque carefully and fire it. This time fast and high and to maturity — red glowing. It will be a breathtaking experience when you take the piece out of the kiln. The figure seems to breathe, the shades and colors have blended. Only the surface is much too glossy and still uneven.

Suppose you are working on a creature which should have some gentle pink hues somewhere. These pinks must never be applied before the very last fire, together with the flux. If the transparent pink, as light as thought, gets several fires, it will be the orangy-red of an apple. Therefore, when wetpacking the last coat of flux, blend the pink into it.

All stoning of the plaque must be done by hand. Take care not to produce deep scratches. While stoning, the artist has his chance to bring out just the right highlights (stoning down to the opaque), to bring certain sprinkled-in opaques to the surface (see discussion of the triptych above), and to enhance transparents which might have become too dark under too thick a coating.

Use great care in stoning, not to stone too deep, to lose the shadings, or to distort the image.

When the surface is finished with scotchstone and wet-and-dry emery paper and the dust is washed off, wax and polish the piece with the warm inside of your hand. There is no better way. Waxing is also the best way to clean and rejuvenate enamels made in this manner.

A few of many drawings made in developing the designs for five panels, starting quite naturalistically and slowly achieving the one line which is concentrated expression.

ENAMELED GOLD FRAME FOR SMALL GOLD CLOISONNÉ PLAQUE

The construction and assemblage of this frame show one way of setting a precious, very small enamel. In the enlargement on page 112 you may see how the design has been translated into a few feet of gold wire and onto less than two square inches.

The unit of letters on the back of the frame is often read as AVE, but it is really an ancient, Romanesque symbol for "Mary." It is composed of the letters: A M A T E, meaning "Love one another." Because of its beauty and decorative value I include it in this handbook as an invitation to craftsmen to use it.

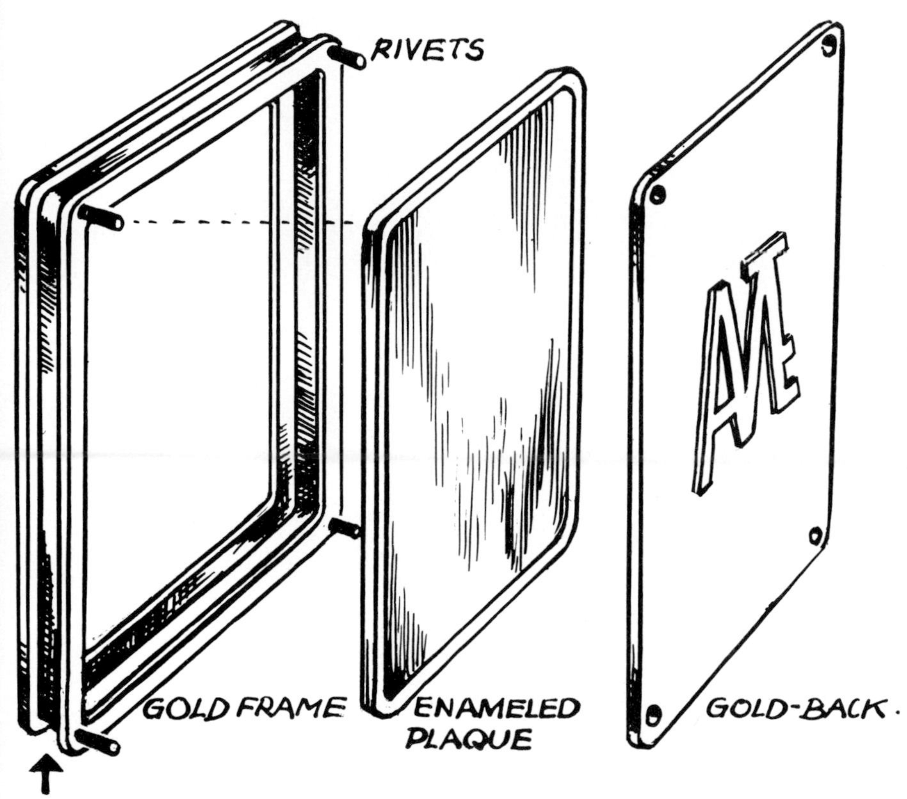

Construction of the frame.

The back of the finished plaque. (The plaque is shown on page 112.)

The American eagle on this gold cup is enameled in gold cloisonné right onto the foot, flush with the surface. A piece of 20-gauge non-tarnishing 18-carat gold was soldered behind the opening which was sawed out for the design. To keep the shape of the foot during the five necessary firings, it was fired over a fireproof cone made to fit the shape. (Courtesy of Mr. R. McConnell)

Two Kiddush cups of 18-carat gold. The cup on the left has an enameled sphere decorated with tiny gold grains (see page 45). Engraved around the equator of the sphere and on the foot are these words from Psalms: "The heavens declare the glory of God; and the firmament showeth His handiwork." (Courtesy Mr. E. Strasser)

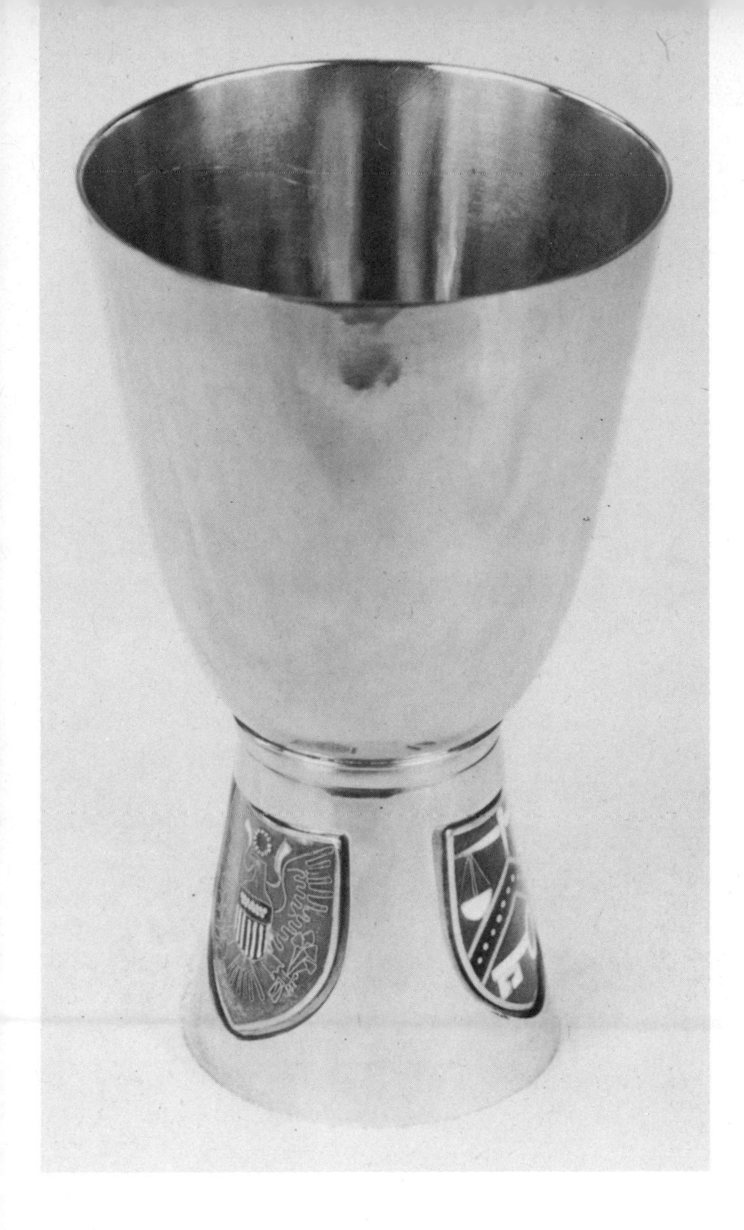

Three coats of arms are fired separately and set into the foot of this cup. Each of the three enamels is executed on non-tarnishing gold with fine gold cloisonné. To enhance the jewel-like effect, the metal was engraved before enameling. After being stoned, the enamels were flash-fired. (Courtesy of Mr. R. McConnell)

The Berlin Cup. 18 carat gold champlevé and gold cloisonné wires. This was made in memory of the end of the Second World War and the rebuilding of Germany. The Berlin bear is shown under a dead tree from which a new green twig is sprouting.

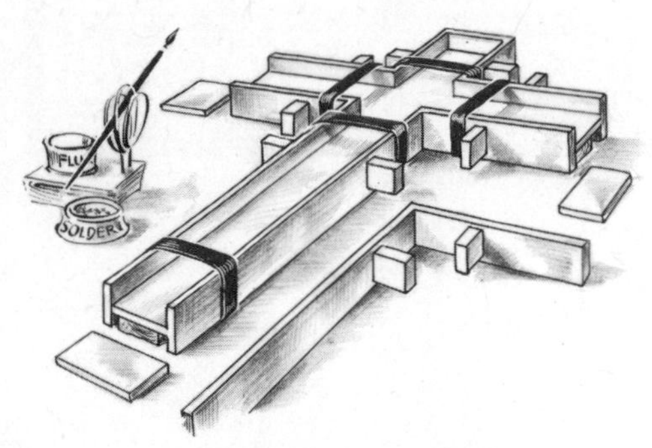

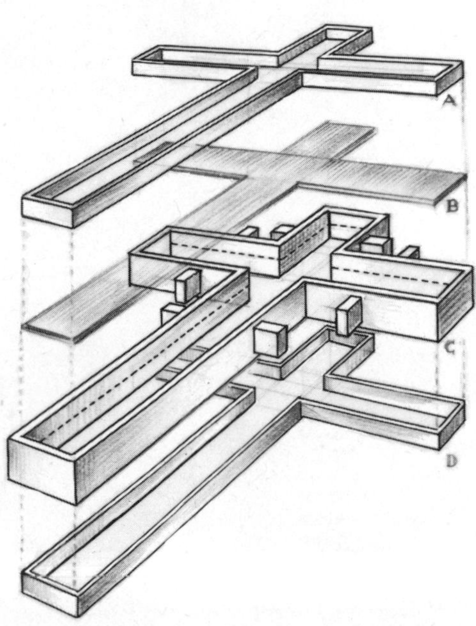

The steps in making a pewter setting for the processional cross.

The complete cross assembled.

PEWTER SETTING OF ENAMELED CRUCIFIX FOR A PROCESSIONAL CROSS

1. Four angles of 8-gauge pewter are made; the ends about $\frac{1}{4}$ of an inch longer than needed. Eight small blocks to be used later for screwing this setting into a bronze cross are made, and prepared with holes the same diameter as the screws to be used, drilled very straight. Only the beginning of the thread is cut. The screws will cut their own thread into the pewter when the cross is assembled. The eight blocks are soldered to the four angles.

With strong rubberbands, the four angles are set and held around the enamel, which rests on wooden blocks $\frac{1}{4}$ of an inch thick. The four small endpieces, which are prebuffed, are now fitted to the four angles; pieces of cardboard between pewter and enamel provide for an easy slip-in.

2. Two cross-shaped frames of 14-gauge pewter are made to fit exactly into the pewter cross. The enamel will be held between these two frames. They extend about $\frac{1}{16}$ of an inch above the pewter cross, to hold the solder. The enamel is protected with cardboard and a piece of sheet asbestos while the frames are tacked to the cross. The soldering of the frame in front is done without the enamel. The front can now be filed and finished. The back has to be completed with the enamel in the cross.

3. The crucifix is copper cloisonné on copper, without flux. The wires, 1 mm. high and 0.20 mm. thick, are set with Klyrfire directly on the bare metal. A rather dark grayish-tan opaque enamel builds the first, relief-like layer of the body. The background consists of a first coat of dark transparent gray, covered in the next fire with transparent reds into which some rough grains of opaque lighter red are sprinkled. Inside the letters I N R I, a roughly ground light opaque red is fired, but it must not be too thick. The dark background must still be felt, if not seen. The cloak over Christ's shoulder is a light opaque red, covered with opal gray. When stoned, some of the red came to the surface. The cloth around the hip is deep bluish gray in many shades, achieved by first firing a mixture of opaque white, gray, and tan, which is heavily covered with several layers of transparent grays. The marks on hands and feet are opaque red inside the wire, blending dark red into the background outside the small wire frames. The shading of the body is done with a greenish-gray transparent (Thomas C. Thompson, number 129), and in a very few areas with a transparent brown. To keep the subdued harmony of color, the stoning was done with great care so that only a few highlights came close to the surface and none up to the surface.

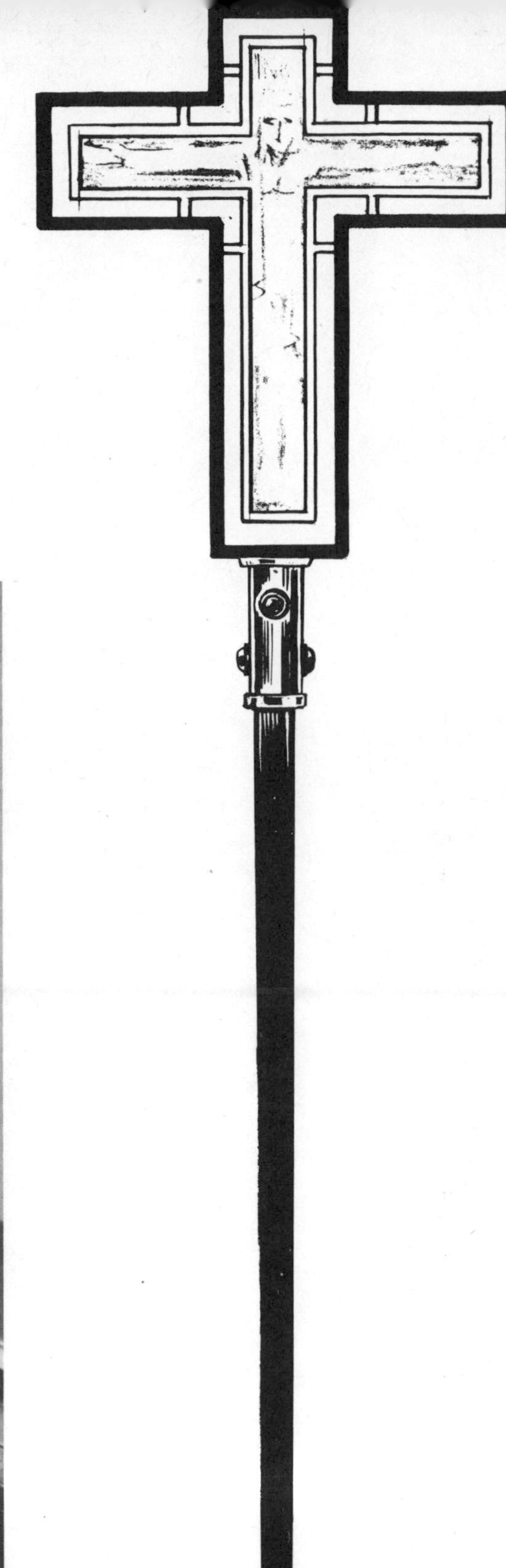

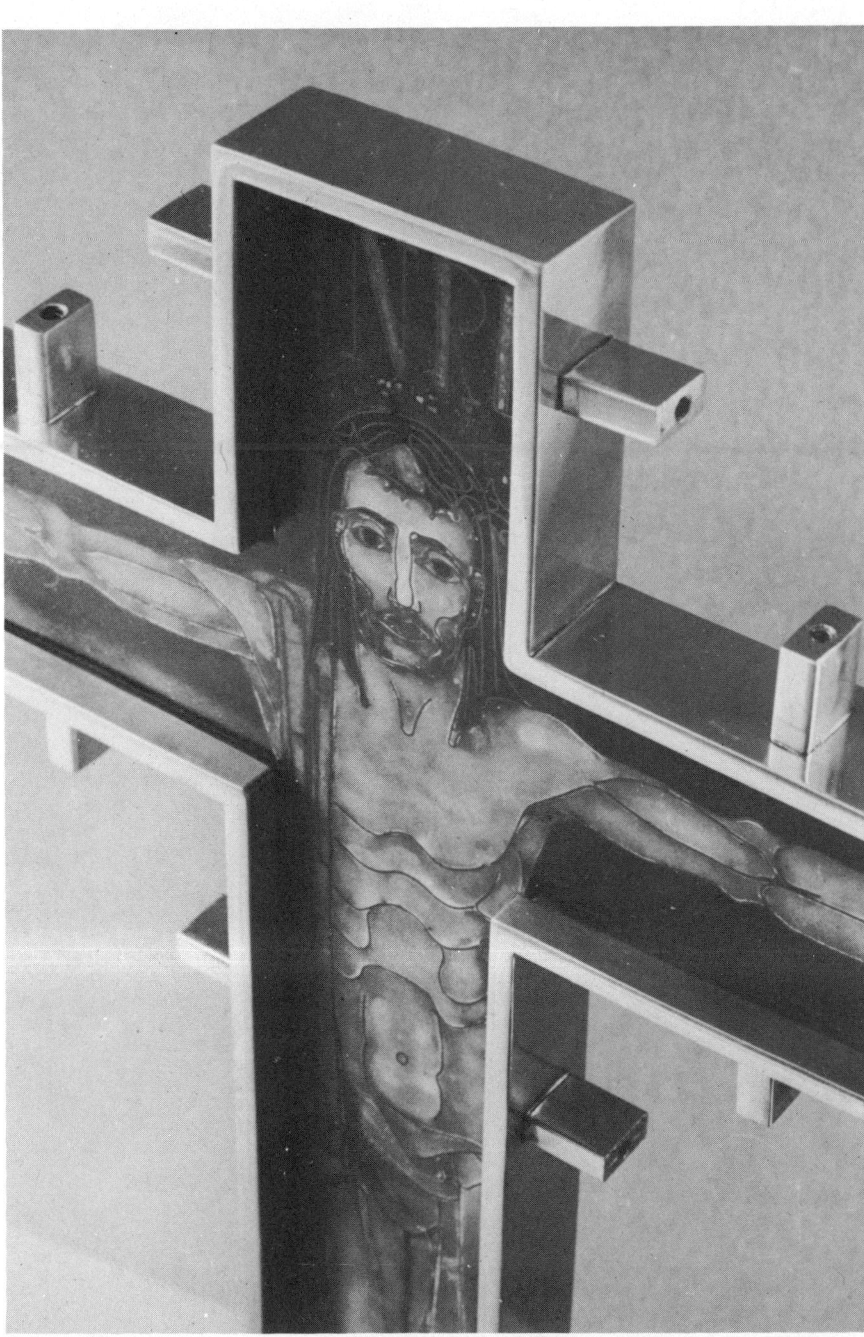

Detail of enamel in the processional cross.

MOSAIC OF GOLD SHEET UNDER TRANSPARENT ENAMEL

Very thin rolled, non-tarnishing 18 carat gold, fine gold or fine silver cut into small shapes may be fired onto the first coat of flux or enamel. While foil will wrinkle, these small shapes remain flat. The enamel underneath must not be thick or they may tilt. To enrich the design, the small gold shapes can be chased with fine lines before placing the first coat of enamel. The gold, being annealed, is so soft that a dull point and very little pressure will do. Place the gold on a piece of leather for a softer and deeper incision. But do not have too soft a background, for the metal will then curl up. (It can easily be flattened again, though.) These lines will remain visible under transparents.

After this chasing is done, the small shapes should be annealed again. To be placed on a sheet of mica and inserted into the kiln for a moment and cooled in water are all that is necessary since these metals do not tarnish. If this technique is used in combination with cloisonné, wires should not rest on the metal shapes but on the coat of enamel. They cannot adhere otherwise.

To surround such small precious shapes with cloisons and fill them with clear brilliant transparents while the background is a dark and even opaque creates delightful effects. There are limits to the size of the small pieces, if they are not pricked with tiny holes, but I found it possible to cover large areas when I used them like a mosaic. (See p. 110.)

Non-tarnishing gold sheet is now available and suppliers are listed in the Appendix.

Chasing small shapes of gold sheet.

Foot of a cross in Washington, Connecticut. A beautifully executed golden snake bites the heel of man.

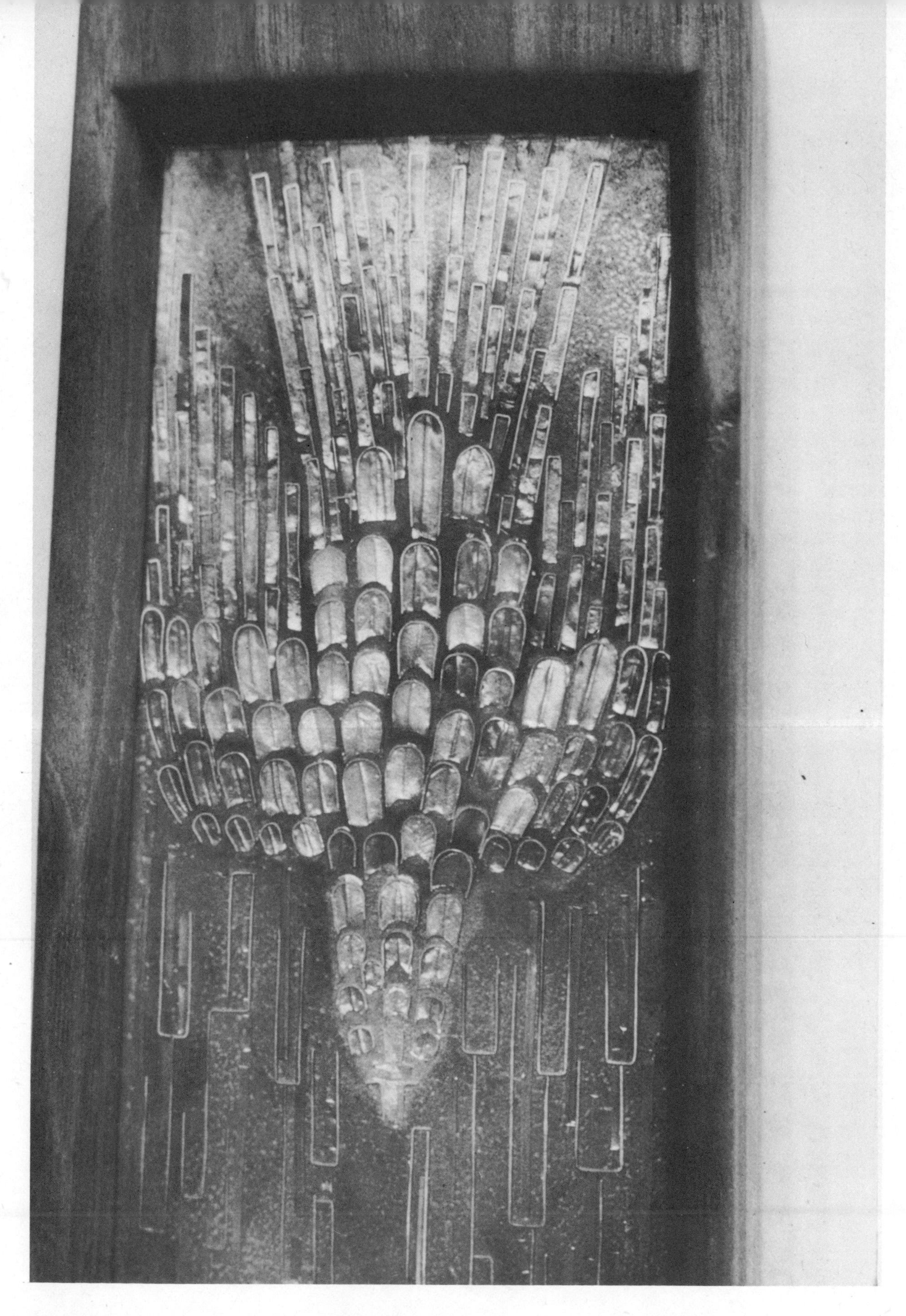

Mosaic of gold sheet, fired under gold cloisonné enamel. Detail of cross in Washington, Connecticut.

A contemporary icon. This plaque is silver cloisonné enamel on copper. It shows how lettering can be part of the design. The theme is misuse of knowledge and science. The hand of the creature representing evil points to a symbol which represents the hydrogen atom.

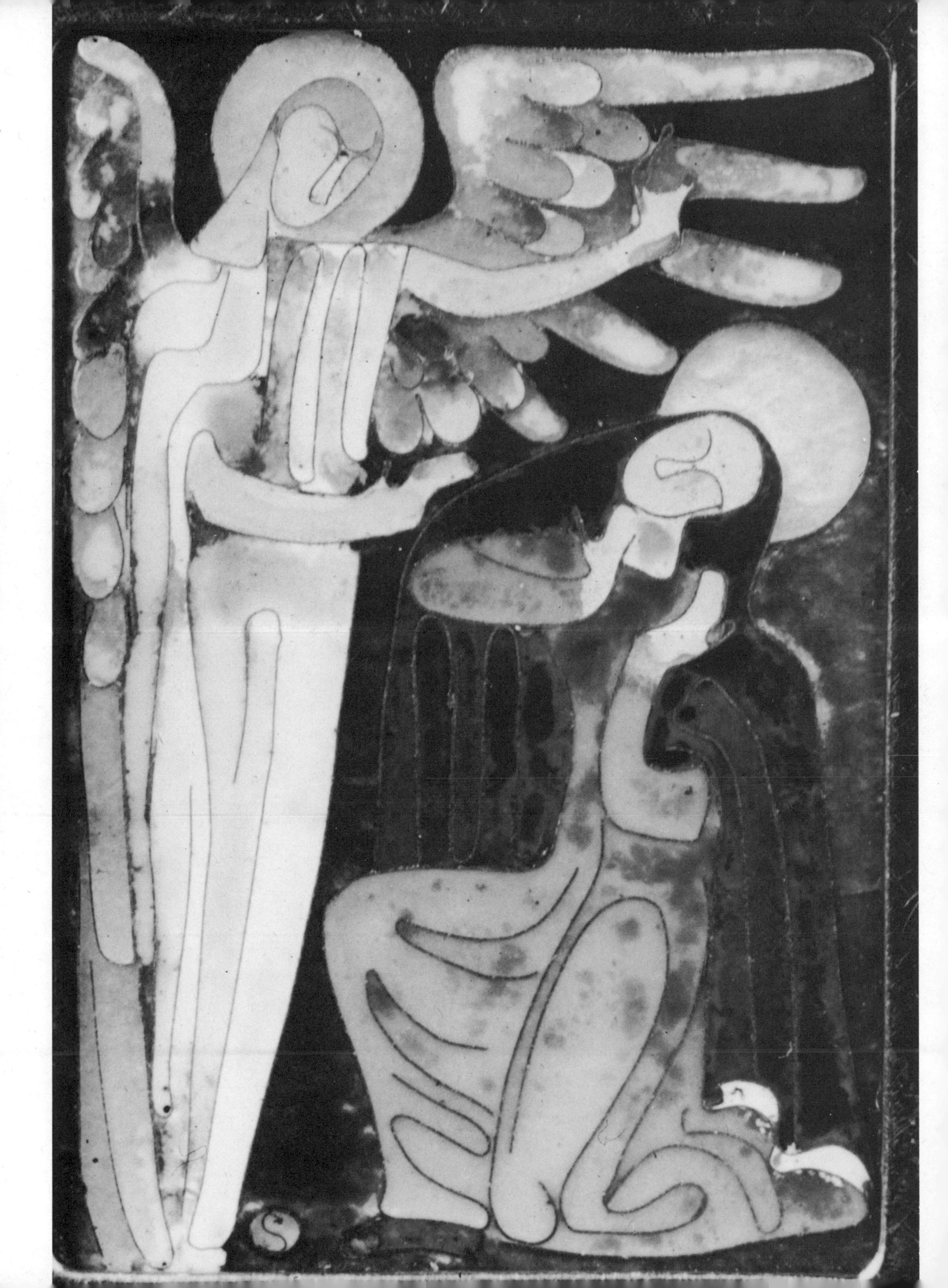

ENAMELING PORTRAIT MINIATURES THREE DIFFERENT WAYS

Though the enameling of portrait miniatures exceeds the scope of this book, I include the recipes from my notebook for the grisaille portrait on p. 73, for the small self-portrait in grisaille, on p. 8, and for the portrait in the pewter frame on p. 46. The latter was done in colors.

1. *Albert Bauer, a medallion in grisaille enamel with fine gold lettering:*

Metal: 16 ga. copper.

No flux. Two evenly sifted coats of (washed and dried) dark gray (No. 122 from Vienna).

Painted with grisaille white, mixed with oils, details very accentuated.

Fired after careful drying.

Second coat in the same manner.

Clear flux (No. 101 from Vienna) in three extremely thin, veil-like layers, sifted only over the face. All surplus around face is removed with a paintbrush. Firing between each coat, but only the last fire to maturity. These three coats of flux change the painted grisaille to the texture of an old man's skin.

Last touches and highlights with white grisaille; lettering with fine gold. A last, careful fire.

2. *Self-portrait, done in two mirrors:*

Metal: 18 ga. copper.

No flux. Two sifted coats of brilliant orange-red transparent (No. 1402 from Vienna), which fires to a brown.

Painted in the same manner as the Albert Bauer medallion, but with no coats of flux.

Last touches with white and fine gold.

3. *Portrait in Pewter Frame:*

From the original a small copy was made which reduces the colors to the shades of background, skin, hair, eyes, cloth.

Metal: 18 ga. copper.

No flux. Two very even coats of flesh-colored enamel were fired over the whole plaque; then it was stoned to perfection, cleaned, and refired.

The detailed drawing was added with painting enamels, prepared with oils. The shadows were brushed in very gently. A dark brown-gray was used; then careful drying and very careful firing so as not to lose the design.

With "water" — enamel: the background, the colors of hair, eyes, cloth were wetpacked, and the whites in eyes (opal) and collar (opaque).

Fire, but not to maturity. Careful firing is necessary to preserve the sensitive painted enamel, which otherwise sinks and disappears in the opaques.

Two separate coats, each as thin as a veil, of clear flux (No. 101 from Vienna) were sifted over the face only; wipe off surplus. (These two transparent coats blend design and enamel beautifully.)

Second coat of background enamel, sifted to achieve better evenness, and addition of "water"-enamel wherever it might be needed. Care must be taken that the dry and the wet applied enamel do not touch. (Water rims.)

Fired to an even surface.

Last touches with painting enamel (some contours were almost lost). White highlights in eyes with a bit of grisaille, and a last careful fire.

The Annunciation. Gold cloisonné.

Two open rings. First pieces of beginners. Silver cloisonné.

Three first pieces by beginners.

Gold cloisonné by an advanced student inspired by an antique mosaic in Israel.

The inside of the pendant is enameled with transparents. By an advanced student.

Gold cloisonné. First ring by a beginner.

Band ring. Silver and enamel by a beginner who is a mason.

Gold cloisonné fish set in a silver pin like a precious stone. Small gold grains complement the design. By an advanced student. Enlarged, original length 1½ inches.

First silver cloisonné by a beginner.

Enamel for a small icon, with gold fired under cross. First enamel of a beginner.

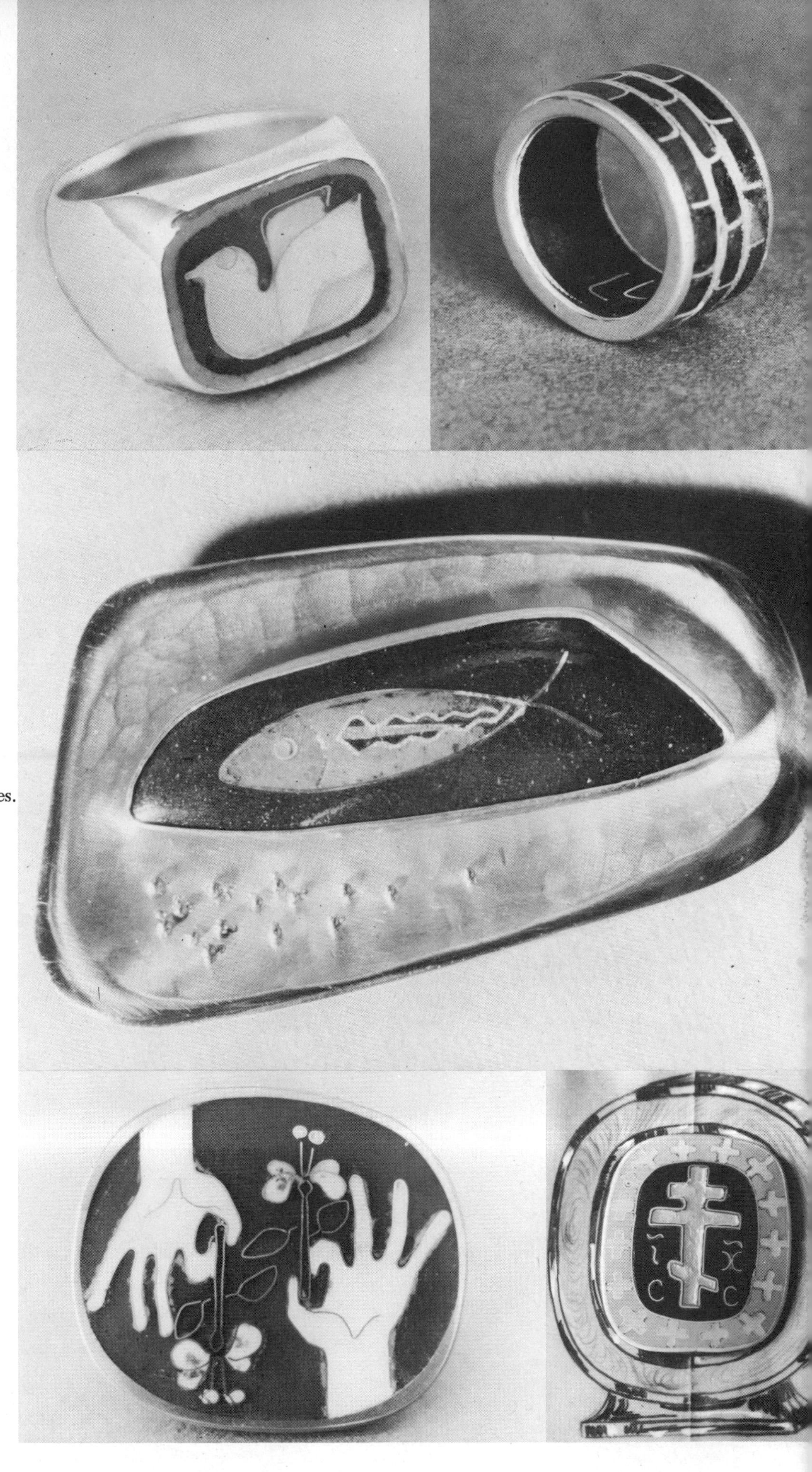

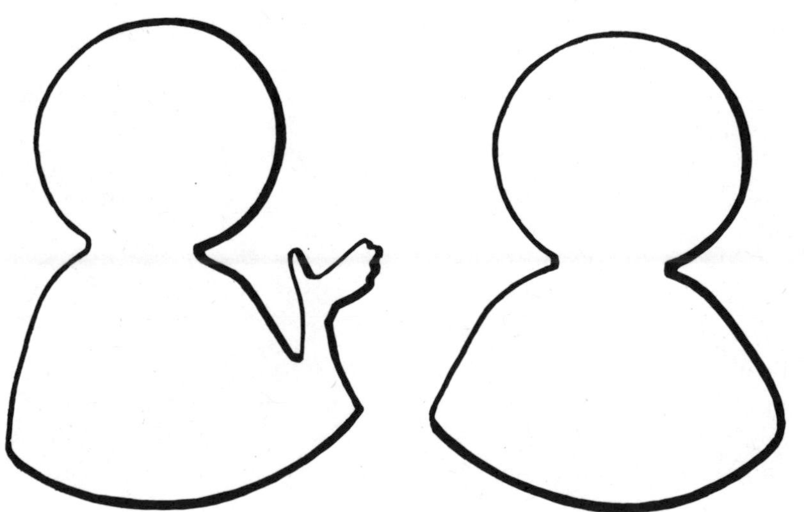

The line silhouettes show the outline of two of the four Byzantine gold cloisonné enamels above. It is possible to saw such outlines from bronze sheet, about 18 ga. thick, and get a clear, deep imprint on fine gold sheet, using both parts of the bronze, one as the stamp, the other as the die (as shown by the diagram). This primitive die will serve two ways: the hand to the right and to the left.

7.

Thoughts on Medieval Enamels

Whoever becomes intrigued with enameling will sooner or later find himself in a museum face to face with precious, skillfully done enamels a thousand years old and still well enough preserved to make him silent with admiration.

We look at the gold cloisonné of Byzantium and think: nobody can do this in our times. Nobody does. Enameled on high carat, thin gold sheet, the metal base is ideal. It is soft and pliable and takes great heat. It has no solder-joints, does not tarnish, or produce cinder. The enamel is protected by innumerable cloison wires placed in a masterly design which makes these miniatures great. There is dignity and lively expression in less space than an ordinary postage stamp occupies.

Still, we know that these enamels were not one-of-a-kind originals by one master, but were manufactured and exported to monasteries and courts as far as the limits of the Christian world. They were treasured and certainly well studied by craftsmen in other less fortunate countries without access to gold and to the last hues of the antique culture.

Byzantium was supplied with gold from the ancient goldmines in the southwestern slopes of the Caucasus. The monks and craftsmen of southern France and along the Rhine had some gold, at least enough to give their work a good coating, but with neither skill nor material could they compete with the Eastern Empire. However, they tried and they had the desire to create objects as beautiful as these precious ones. Copper they had, and enamel in colors as radiant as those from the East. And they had the genius to do it in their very own way. Instead of bending the thousands of minute gold wires, they prepared the metal for champlevé, reducing the design to fewer strong lines. The approach of these medieval craftsmen can teach us much. Notice how they nailed the enameled, mercury-gilded plaques on a small wooden reliquary. Nothing is hidden and yet it is honest and good craftsmanship. The rather primitively designed figures have meaning and style. The colors are nearly the same as these in the small gold cloisonné plaques from Byzantium, which one might compare with a seed that fell on fertile ground and something new and strong grew from it.

How may the workshops in Byzantium have proceeded with their production? I do not say that this is how they did it, but let us try to find out how it could be done. Obviously, the very compact contours of the figures are set deep with a stamper (maybe of hardwood) which can easily be made and works well on soft gold sheet. (It can also be done with thin copper, but we have to watch for cinder.) If such a stamper is sawn out from 18 gauge copper, we automatically have a negative die. The depth of the impression need be no more than 0.7 to 0.8 mm. The few letters may have been imprinted in the same manner.

Studying the four reproductions on page 116, you will see that 1 and 2 have exactly the same contour and so have 3 and 4. Only what was left became right and right became left. Skilled hands could reproduce what had been done once. The beaded rims, chased and perhaps also prestamped from the back, provide stability and a good enclosure. There is system, and one might almost call it standardization, in the bent wires. This means that the small units, filling larger areas, could be prebent by skilled hands and be available to the man who assembled the design. Nothing is wrong with this. I admire the high quality of every single finished piece which has left these old workshops.

I am sorry that I cannot answer the question as to whether or not the antique pieces are counterenameled: I have tried to get the answer, but it was not granted to me to look at the back of an old enamel. For our own work I suggest counterenameling. When I experimented with similar techniques, the enameling held no problems at all (I speak of the technique only, not of the great art the old pieces represent). But when stoning, there is danger of stoning through the metal at the contour. The craftsmen of old had similar problems, as their work shows.

I include these thoughts about medieval enamels in the hope that some fine artist-craftsmen will pick up from here and create beautiful things, their own in conception but technically based on the knowledge and skill of the old masters.

A twelfth-century French box. Champlevé enamel on copper and mercury gilt. The six plaques are nailed onto the wooden coffer.

Epilogue

Many pages in this book show enlarged details of minute work and enamels which have been executed with very different approaches so that you may study them closely. The cloisonné work reproduced invites the interested craftsman to follow the sections of wires — perhaps even with a pencil on a piece of transparent paper — to get an idea of what can be done with a few inches of wire. What at first glance seems to be a rather naturalistic representation is really concentration on a single line, characteristic not only for its outward shape but for the feeling it expresses.

Some of my plaques are like contemporary icons. I did not intend to illlustrate the Bible, but its figures represent the eternal challenges. I am also very fond of Greek mythology. The Greeks had a way of symbolizing the superhuman and the most human. And there is that Homeric laughter . . .

Sometimes my work is just relaxed playing with wires on a little box, which in itself is a plaything. Sometimes the work is naive, sometimes comparable to rope-walking.

We have come to the end of this book; I do hope that on the foundation laid here you will go on by yourself and be rewarded with success as you attempt more and more difficult work.

Sentimental Centaur. Gold-enameled box. (From the collection of Robert Laurer)

Appendix

Detail of Eternal Light. *Plique à jour* in fine silver.

TRANSFERRING DESIGN TO METAL OR ENAMEL

No carbon paper, please; it is greasy and does not disappear when fired. If your design is on tracing paper, which I very much recommend, lay the paper over the metal and prick certain places with a sharp scriber through the paper into the metal. From these places you can start to set cloisons or draw with Underglaze D.

Or, if a detailed design is needed, paint the metal with a coat of tempera white. Let it dry and place graphite-coated paper and the drawing over it. Draw the lines with a hard pencil or dull point, as you would work with carbon paper. If graphite paper is not available, rub the back of your drawing with a very soft pencil, place it over the white-coated metal and trace the lines with the hard pencil. After removing the tracing paper, incise the lines with a sharply pointed scriber through the white coating onto the metal. Wash the white color off, and you can start enameling.

If a coat of enamel is already fired: If the enamel is to be like a jewel, nothing should be applied which might leave traces. Hold the design on transparent paper carefully over the enamel and check where you might start, making some pencil marks which can be wiped off later on. If you work on a large plaque which is not too sensitive, coat the design with stabilo pencil, quite thickly applied, trace it with a hard pencil, and either go on enameling or draw the lines with "fine line black" (used on enamel, not under it like Underglaze *D*). Fire them and you have a permanent design on the enamel.

Models should be made of important work. If it is three-dimensional, use wood or cardboard, painted in the intended colors.

If it is a panel, draw the lines of cloisonné on a pane of glass, tracing them from the back of the transparent paper, which is the original design. Then make a separate color-background, using gold-paper, paints, whatever comes closest to the hues to be achieved, glued on cardboard in the manner of a collage. Lay the pane of glass with the design over this improvised color scheme. Since design and colors are separated, experimenting and changing are easy. The shades the artist decides upon will be found on his color sample plaque and he needs only to read from it to achieve the right color.

WHEN TO COUNTERENAMEL

After the front has one coat of enamel fired to it, or before the front is done?

Copper: If the front is to be very clear, I like to take into account the *doming* which takes place by itself when you enamel the *front first.* The back is protected by Scalex. The counterenamel is applied second.

On sterling silver: First a neat counterenamel with *transparents* which look good directly over silver is advisable. Pickle the piece, engrave or finish the front in order to enhance reflection and to give more hold to the enamel.

Sterling casts: Enamel the back first. No oxidation appears on the front as long as the fine silver-coating is not damaged.

Fine silver cannot oxidize. Counterenamel first so that you don't have to be concerned with it any more, then proceed with the front.

Gold: The back should be treated with the same care as the front. Enameling the back first makes work on the front much easier.

Band rings, either gold or silver or both metals: Enamel both sides at the same time, using suitable transparents.

Tall, vertical shapes (foot of a chalice): Outside first, using flux, since it is elastic and will not crack. If a dark color is desired, mix the flux with this color (1 : 1). Use Scalex on the inside if the shape is copper. Fire and, after it cools, clean and enamel the inside. Places that cannot be reached with the sieves are "salted" by hand. Be sure that enough Klyrfire has been sprayed, and respray if necessary to hold the counterenamel solidly. It is possible to fire this counterenamel simultaneously with the front.

Boxes: If possible, spray Klyrfire and coat the interior and the exterior at the same time. If this is not possible, coat the inside first. Then go on with the outside.

ACIDS AND ASPHALTUM

Proportions of Solutions

Always add acid to water, never water to acid.

Nitric acid is used for cleaning gold, to enhance its color — and for etching copper and silver. Nitric acid does harm silver and copper, so never immerse either metal, nor anything soldered with silver-solder. A few drops added to water when enamels are washed cleans the colors and removes impurities.

Nitric acid

3 parts water
1 part acid

Sulfuric acid is pickle for cleaning gold, silver, copper, or Tombac. Never permit any iron to get into this pickle; use copper tongs to remove pieces from the acid bath. No binding wire should be left on objects to be pickled. If it does happen, silver and gold will be coated with copper. This is very hard to remove. You have to make new pickle.

Sulfuric acid

10 parts water
1 part acid

Hydrofluoric acid, the most dangerous of acids, must be stored either in a lead, rubber, or special plastic bottle. It destroys glass, enamel, and causes bad burns on the skin. The fumes are poisonous and the jar into which objects are immersed should have a cover and be made of *acid-resistant plastic.* Even so, I coat the inside and the cover with wax. (Heat the wax in a double-boiler and slosh the liquid wax around in the bowl.) A white layer of dissolved glass appears where the acid has worked; this should be wiped off with a feather or a cotton swab, since

Hydrofluoric acid

1 part water
1 part acid
or
2 parts water
1 part acid

this layer retards the etching process. Only a coat of wax will protect the areas not to be etched. Painting the acid over a glossy surface will dull that surface. Warm acid works faster than the cool liquid. It is used to remove enamel and does not attack gold, silver, or copper.

Asphaltum prevents etching with nitric and sulfuric acids when it is painted as a thick coat over the metal. "Arnesto" asphaltum stays pliable enough to permit incisions through it, after the surface is just dry. If the asphaltum is too dry, it would chip under the dull metal-point used as a scribe when designing an etched pattern. The asphaltum is ready for the design about one hour after it is painted over the metal. Paint asphaltum thick. No specks of dust or thinly covered areas.

Asphaltum is removed with unleaded gasoline. It is not as greasy as kerosene nor as sticky as turpentine.

White opaques are sensitive to acid; check other opaques if necessary.

There comes a time when a stoned enamel should not be fired again. If part of the surface is still glossy, paint hydrofluoric acid over it; this will dull the surface. The enamel can then be waxed and hand-rubbed.

If it is too risky to clean an enamel in pickle, you can put the plaque on a piece of aluminum and immerse it in a warm soda solution. Or, of course, you can clean it mechanically.

APPROXIMATE GAUGE AND MILLIMETER EQUIVALENTS

1.00 mm. =	18 ga. =	.0403 inches		
0.80	20	.0319		
0.75	21	.0284		
0.50	24	.0201		
0.25	30	.0100	Heavy	
0.20	32	.0079	Medium	Silver Cloisonné Wire (18 ga. high)
0.15	35	.0056	Fine	
0.10	38	.0039		Gold Cloisonné Wire (20 ga. high)
0.07	41	.0028		
0.05	44	.0019	Hair-fine	

USEFUL HINTS AND SUGGESTIONS FOR REPAIRING ENAMELS

To start this frightening yet comforting section with an example of a typical last-minute catastrophe, let me show how an invisible repair was made. It is a true story, and it happened to me.

On a complicated, finished piece, which was stoned and polished and ready for the very last flash-firing, a big bubble appeared under a gold-cloisonné letter. The wire stood almost vertical; the bubble was very deep and quite discouraging.

Using a diamond-impregnated small wheel in the flexible shaft, I ground out the wire and the entire affected area, going as deep into the enamel as the weakness seemed to be. These small wheels and points are expensive, but they are worth every penny. You can go very close to the next wire and do as precise a job with the enamel as a good dentist does on teeth.

With glass brush and running water, I cleaned the cavity carefully and filled an even, thin layer of enamel into the bottom, using the identical color. I fired the piece not too high or too long, just enough to get it slightly fused. Then I replaced the cloisonné letter, imbedding it firmly into enamel and adding a drop of Klyrfire to hold it in place. It was surprising how much enamel could be packed into the hole with a small mound over it.

I saw to it that I could reach the "disaster area" readily with a small spatula the instant the piece was taken from the kiln, and I placed the piece on a trivet with this in mind. At the very moment the plaque emerged from the kiln, I pressed the letter down into the enamel gently and quickly, keeping calm. A small depression around the wire remained. When it was stoned smooth, the repair was finished. (If this does not work the first time, refill enamel around the repaired area, fire the piece once more, and then stone and finish it.) Not even I could tell a repair had been made.

Hints for working with metal

Out-of-shape plaques: Either the construction was not right or the trivet did not suffice to hold the plaque. As long as it has only flux, it might be possible to bend it back cold. Otherwise, refire on a flat trivet and press between two sheets of asbestos while still glowing hot.

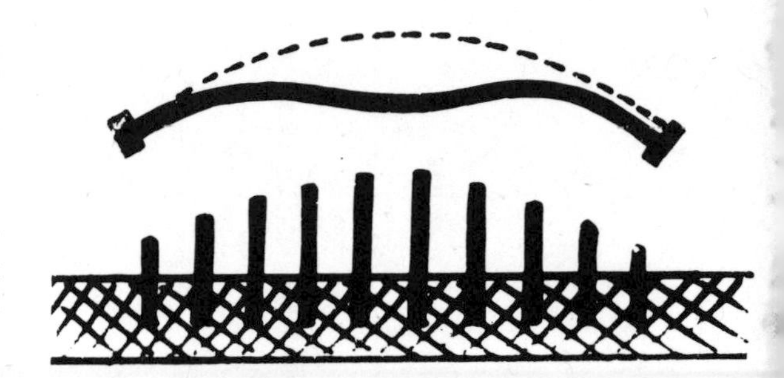

Out-of-shape (concave) panel: Construct a trivet which is taller in the center and refire the panel high enough to cause it to settle back into its intended shape. While it is still hot, try to press the rim evenly flat.

Bowls to be enameled keep better shape and have much finer quality if a ¼ inch rim is sharply turned either in or out. This rim must be hammered tightly to the body of the bowl. Otherwise, enamel might creep underneath or air pockets remain. This rim receives NO enamel and protects the enamel. It is to be buffed and polished and may even be gold-plated.

Hints for working with enamel

When enamel fires porous, either the metal was not a good alloy, or the piece was overheated, or the enamel was not clean and washed, or there is some solder underneath.

Whites turn yellow around the edges due to overfiring.

Causes of cracks: The metal construction may have been inadequate; there is no counterenamel; the piece was under-fired or cooled too fast after firing; or hard enamel was fired over soft enamel. The soldered rim may have become loose, or the cloisonné wires or partitions may not quite touch, leaving open spaces and causing tension in the enamel. Finally, there are colors which do crack, for instance, certain chartreuses.

Surface cracks appear when the last coat of color covers metal partitions. Usually these can be stoned off. If not, they will close in the next firing when metal is free from enamel.

Cracks deep under the surface: Has a hard color been used over a softer enamel? Such a place must be ground out and refilled with soft enamel. Where metals did not quite meet, you can insert small cloisonné shapes; or a gold ball may work sometimes if it is fired deep enough into the enamel. Grind such a gap as deep as the diamond-impregnated wheel can reach, and repair as suggested.

Cracks near rims that were soldered to the vertical shape may be due to the fact that the outer rim was very wide and too much enamel is left where horizontal and vertical meet. Stone the vertical coat evenly and refire.

Check to be sure that the inside is counterenameled well.

If the last coat of enamel is all wrong, cover the back of the panel with melted wax and immerse the piece in hydrofluoric acid. It eats away the enamel in layers; if you watch, you can control the process until the layer you want to eliminate has been destroyed. The cloisonné remains intact. Remove the white residue thoroughly with a brisk brush and plenty of running water. When only the "healthy" enamel is left, wash the piece in ammonia and warm water, rinse, and glass-brush very carefully. Then refill the enamel and fire.

To remove enamel from certain areas use the same procedure as above, but cover all areas to remain unhurt with a good heavy coat of molten wax, for only wax resists this powerful acid.

If the last cover of flux is milky, the flux was too finely ground and not washed well enough, or it was over-fired. To have guaranteed clean flux, grind it freshly and add five drops of nitric acid to the washed material. Stir, and after five minutes pour the water (and acid) away and wash the flux again at least six or eight times. Enamels become harder this way, but the acid removes all impurities.

Repairs without firing: It can happen that a piece cannot be fired again and still needs some small repairs. Perhaps it is an old piece, made by someone else (never heat such an enamel).

There are so-called cold enamels available, evidently some sort of sealing wax, which never seem to be the right color.

If white epoxy is mixed with small quantities of artist's oil colors and the color-matched paste is used as filler, the repair will be as perfect as can be expected. Let it dry at least 24 hours; then wax the surface.

Cracks cannot be concealed in this way, but they can be made less obvious with spermaceti. This is a fatty substance produced by whales; it is available at many drugstores.

Before applying spermaceti, warm the panel carefully. The safest method is to immerse the piece in hot water.

Another repair of a "hopeless case": The solder inside a rim had not been removed and cleaned. In over-firing the enamel became very porous and rough.

The entire margin was removed at a constant width, then washed under running water, and glassbrushed. The same color, but opaque and just a trifle lighter than the original *transparent,* was filled into the margin. After stoning, this new enameled rim was more attractive than the original one!

PIN STEMS

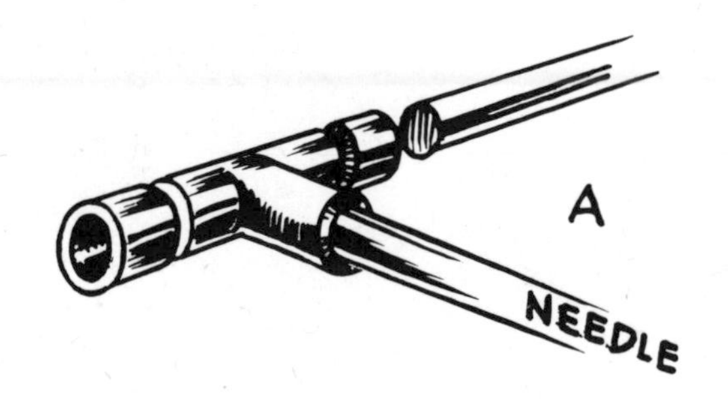

There are good findings available but pin stems seem to cause trouble. Here is a hint on how to make pin stems which will not break or bend. Use 16 gauge metal, a hard alloy of gold. In a silver pin, the sterling silver should not be annealed. Saw out a shape as shown in sketch (A). Hammer it hard without thinning the metal and try to hammer it round, tapering where later the needle point will be filed. Finish and polish this piece, fit it into the hinge, and mark the spot where the hole will have to be drilled. The small tongue (B) provides elasticity. Rivet the needle to the hinge and file the tongue so that it achieves the proper spring-tension. This pin has never been heated and remains hard; it is strong where other pins would break.

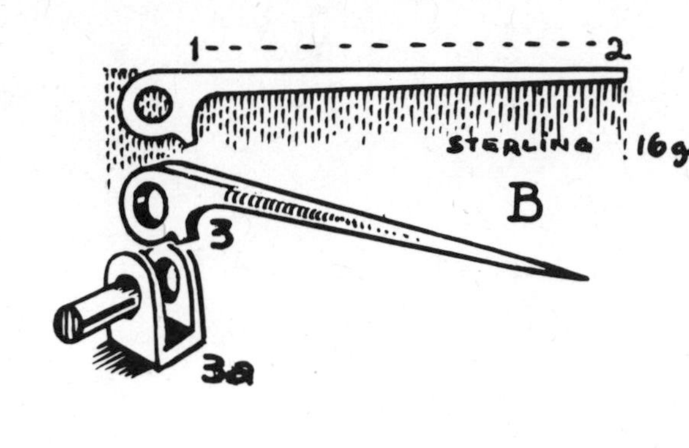

A pin stem which will not damage very fine fabric: The hinge is made of tubing. Solder another small piece of tubing in an angle of 90° to the moving middle section of the hinge. Then hard-solder a sewing needle (without the eye) into this second piece of tubing and attach a small ball underneath this piece to provide elasticity later on.

Findings should be attached to good enameled jewelry with hard-solder whenever possible. When this cannot be done, provide for platforms hard-soldered to the piece to be enameled, and equal in size to the platforms on the finding, for soft-soldering. Melt the solder over the platform of the finding *first*, hold it in place with cross-lock tweezers, and solder carefully. The tweezers press the findings in place; *don't take them away before the joint has cooled.*

LETTERING

Lettering can be an important part of a piece of cloisonné enamel. It can even be made into *the* ornament. Here is a suggestion for bending wires to form the letters of the alphabet. Start at the point and pinch the wire to single lines (indicated by the double lines in the sketches). Except for the X, all the letters are bent from one piece of wire. It is advisable to trim the I a little for safer standing, as shown in the small sketch.

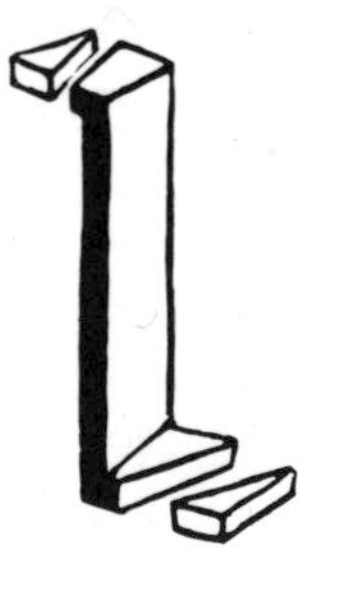

ABCDEFGHI
KLMNOPQRS
TUVWXYZ

• Start here.

▬ Pinch tightly together.

RECOMMENDATIONS

Enamelers who wish to learn more about silversmithing should read: *Handwrought Jewelry* by Lois E. Franke, McKnight & McKnight Publishing Co., Bloomington, Illinois.

I also wish to give credit to the German book, *Metal*, by Wilhelm Braun Feldweg (Otto Maier Verlag, Ravensburg), which taught me mercury gilding.

SUPPLIERS

Non-tarnishing 18 carat gold (N-T gold) will be supplied in sheets and as cloison wire, also as thin-rolled gold to fire under enamels and sponge for painting on enamel by T. B. Hagstoz & Co., 707 Sansom Street, Philadelphia 6, Pa.

Enamel and kilns are available from Thomas C. Thompson, 1539 Deerfield Road, Highland Park, Illinois.

Grisaille white and enamels from Schauer & Co., Vienna - Atzgersdorf, Austria.

Kilns, tools, precast sterling shapes for enameling from Allcraft Tool & Supply Co., Park Avenue, Hicksville, New York.

Gold and silver from Hoover & Strong, 111 Tupper Street, Buffalo 1, New York.

Findings from Allcraft; and Anchor Tool & Supply Co., Inc., 12 John Street, New York, N. Y.

Gems from International Gem Company, 15 Maiden Lane, New York 7, N. Y.

Wet-and-dry paper, asbestos stove-lining available from hardware stores.

Index